T0140371

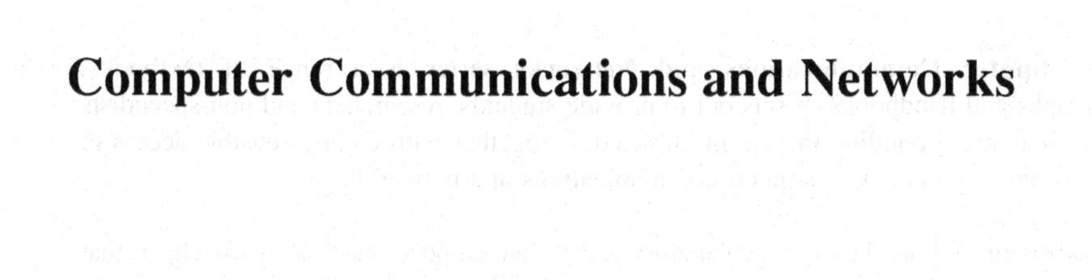

Computer Communications and Networks

For further volumes:
http://www.springer.com/series/4198

The Computer Communications and Networks series is a range of textbooks, monographs and handbooks. It sets out to provide students, researchers and non-specialists alike with a sure grounding in current knowledge, together with comprehensible access to the latest developments in computer communications and networking.

Emphasis is placed on clear and explanatory styles that support a tutorial approach, so that even the most complex of topics is presented in a lucid and intelligible manner.

José Cecílio • Pedro Furtado

Wireless Sensors in Industrial Time-Critical Environments

José Cecílio
Department of Computer Engineering
University of Coimbra
Coimbra, Portugal

Pedro Furtado
Department of Computer Engineering
University of Coimbra
Coimbra, Portugal

Series Editor
A.J. Sammes
Centre for Forensic Computing
Cranfield University
Shrivenham campus
Swindon, UK

ISSN 1617-7975
ISBN 978-3-319-37854-1 ISBN 978-3-319-02889-7 (eBook)
DOI 10.1007/978-3-319-02889-7
Springer Cham Heidelberg New York Dordrecht London

Softcover reprint of the hardcover 1st edition 2014

Printed on acid-free paper

Springer is part of Springer Science+Business Media (www.springer.com)

Preface

Large industrial environments need a considerable degree of automation. Distributed control systems (DCS) have to constantly monitor, adapt, and inform humans concerning process variables. Humans on their side take actions and command modifications to physical variables through the control system. Many of the application scenarios have safety critical requirements, with timing bounds and fault tolerance as important ones. This has led to the development and commercialization of various protocols, standards, and systems.

Some years ago, a new revolution started with the introduction of wireless communications in very different settings, including industrial control systems. Wireless communications reduce the amount of expensive wiring, but for their wide use in industrial settings and safety critical environments in general, it is necessary to provide reliability and timing guarantees.

The objectives of this book are to introduce the components, operations, industry protocols, and standards of DCS and to show how these can be designed with wireless parts while at the same time guaranteeing desired operation characteristics. It contains not only theoretical material but also results and lessons learned from our practical experience in implementing and testing those systems in a specific industrial setting – an oil refinery.

The book is intended for a wide audience, including engineers working in industry, scholars, students, and researchers. It can serve as a support bibliography for both undergraduate and graduate level courses containing the themes of industrial systems, networks, real-time systems, wireless sensor networks, or embedded systems.

Coimbra, Portugal José Cecílio
Coimbra, Portugal Pedro Furtado

Contents

List of Acronyms

ACK	Acknowledgment
ADCs	Analog-to-Digital Converters
API	Application Programming Interface
CPLD	Complex Programmable Logic
CPU	Central Processing Unit
CSMA	Carrier Sense Multiple Access
DACs	Digital-to-Analog Converters
DCS	Distributed control system
FDMA	Frequency Division Multiple Access
FPGA	Field Programmable Gate Array
I/O	Input/Output
ID	Identification
IP	Internet Protocol
LAN	Local area network
MAC	Medium access control
MCU	Microcontroller Unit
OS	Operating system
PC	Personal computer
PID	Proportional–Integral–Derivative controller
PLC	Programmable Logic Controller
PLCs	Programmable Logic Controllers
QoS	Quality of Service
SCADA	Supervisory Control and Data Acquisition
SQL	Structured Query Language
SYNC	Synchronization
TCP	Transmission Control Protocol

TDMA	Time-division multiple access
UDP	User Datagram Protocol
WSAN	Wireless sensor and actuator network
WSN	Wireless sensor network
WSNs	Wireless sensor networks
XML	Extensible Markup Language

List of Figures

List of Tables

Chapter 1
Introduction

Distributed control systems (DCS), also known as networked control systems (NCS), are present in many different environments. For instance, industrial premises, such as an oil refinery, have hundreds or thousands of sensors and actuators, which automate monitoring and control. The DCS is the distributed system that connects and controls all those parts. The energy grid is also controlled by a huge decentralized Supervisory Control and Data Acquisition (SCADA) system that reaches every single energy distribution outpost, constantly collecting measurements and controlling energy distribution along the electricity highways. The purpose of those systems is to automate supervision and control of distributed cyber-physical systems. There are many different data reading sources (sensors), designed to capture physical variables such as flows, pressure, temperature, or quantity of electricity, and many different actuators, designed to change the state of physical parts in response to commands. The DCS constantly monitors the physical systems and transmits actuations and controls by adjusting physical parameters in response to state changes. It also informs humans and allows them to take action.

In many industrial sites individual sensors and actuators of the DCS are connected to some intermediate equipment through a huge web of cables spanning many kilometers and costing a fortune. About a decade ago, industry suppliers have started deploying wireless sensor and actuation solutions, which are easier to deploy and less costly than totally cabled ones. These solutions are based on small embedded devices that are able to sense and actuate, as well as communicate and compute. The promise is that, if wireless solutions are proved reliable enough, they can revolutionize critical applications by allowing sensing and actuation at significantly lower costs.

The distributed control system (DCS) is usually organized as a hierarchical network, consisting of heterogeneous devices with different functionalities. At the very ends there are sensors and wireless embedded devices, which are connected by cables or wireless to Programmable Logic Controllers (PLCs) that link and control field devices. At the other end, there are computers acting as control stations and other parts of enterprise information systems interacting with the DCS. The inclusion of wireless sensors contributes to the heterogeneity of the whole distributed system.

J. Cecílio and P. Furtado, *Wireless Sensors in Industrial Time-Critical Environments*, Computer Communications and Networks, DOI 10.1007/978-3-319-02889-7_1,

Typically, multiple wireless sensor devices will be organized into wireless sensor networks (WSN) spanning whole sensing regions. The wireless sensor networks are different from conventional distributed systems built with computer nodes in many respects. Resource scarceness is the primary concern, which should be carefully taken into account when designing software for those networks. Sensor nodes are often equipped with a limited energy source and a processing unit with a small memory capacity. Additionally, the network bandwidth is much lower than for wired communications, and radio operations are relatively expensive compared to pure computation in terms of battery consumption, if the wireless device is powered by battery.

In the context of DCS, operation timings are very important. Timing requirements can be formulated simply as a maximum time for an operation to take place or some statistical property of time taken by an operation. An example of a timing requirement is that the time from when a temperature passes some admissible threshold and the instant when that event raises a visible alarm in the control room must be less than 1 s, or when an operator presses a command to close a valve, the valve must be closed in less than 1 s. Timing requirements are especially important to meet real-time control needs.

To provide timing guarantees, planning algorithms are needed which will predict operation latencies and dimension the network accordingly. Heterogeneity resulting from wired and wireless parts adds to the problem of how to achieve timing guarantees. Current solutions to plan DCS for operation timings require engineering decisions based mostly on a set of rules of thumb. Consequently, besides introducing the reader to the concepts behind both DCS and associated technologies, this book addresses the issues of how to provide adequate planning for operation timing guarantees in the heterogeneous DCS built with wired parts and wireless sensor subnetworks.

In order to plan for operation timing guarantees, wireless sensor subnetworks can be provisioned with preplanned time-division multiple access (TDMA) protocols. Besides that, there is a need for a model to estimate and control operation timings end-to-end, spanning the whole DCS and considering the wired and wireless paths from the moment an event occurs to the moment it is perceived. The planning phase must decide whether to partition subnetworks in order to meet timing requirements, as expressed by application needs. Consequently, mechanisms are described to plan, dimension, and predict operations latencies in the whole DCS, with predictability and accuracy as objectives.

The book is organized as a set of chapters, each describing one specific issue that leads the reader to better understand the concepts behind heterogeneous distributed control systems, as well as scheduling and planning in those systems, and then how to plan heterogeneous distributed control systems with wireless sensor networks.

Chapter 2 overviews concepts, components, and architectures supporting distributed control systems. It first discusses basic concepts, components, networking technologies, topologies, and transmission mediums. Then it describes how the industrial control systems are built based on those elements.

Chapter 3 takes a closer look at the operations that the DCS implements. It focuses on the software components and functionalities that are implemented in those systems. These include human–machine interfaces, diagnostics and maintenance interfaces, and controllers. Controllers are fundamental software components through which users and automated logic interact with the whole cyber-physical system. The main operations implemented in a DCS are overviewed.

Chapter 4 discusses industrial control system protocols and current solutions for planning those systems. We review the concepts behind the fieldbus and fieldbus follower standards, and we describe the currently available guidelines for planning over fieldbus. Planning is mainly based on guidelines and ad hoc engineering decisions. In Chap. 4 we also review wireless industrial network protocols and the guidelines that can be followed regarding planning for those networks. After this chapter the reader will understand the industrial protocols for both wired and wireless networking and will also understand current solutions to plan operation timings.

After reading Chaps. 2, 3, and 4, the reader will be acquainted with the technologies and components of DCS, the operations that they implement, and the basic guidelines for planning.

The rest of the book describes how to achieve systematic planning for timing guarantees over heterogeneous wired and wireless DCS.

Chapter 5 discusses scheduling in wireless sensor networks. It sheds light into the medium access layer protocols and explains the mechanism of time-division medium access protocols and finally how schedules are achieved within those protocols. The capacity to define schedules in wireless sensor networks is very important for planning operation timings in the whole DCS.

Chapter 6 describes a latency model for heterogeneous DCS with wired and wireless parts. This model predicts monitoring latencies, command latencies, and closed-loop latencies. Based on this model, we describe the approach for prediction of maximum latencies.

Chapter 7 describes how to plan operation timings systematically, and Chap. 8 describes an approach to monitor and debug performance. The planning approach is based on the construction of schedules and the latency model. Based on user inputs and a first network layout, the approach determines maximum WSN latencies first, computes the required number and placement of downstream slots to meet command and closed-loop latencies, and then verifies whether the latency requirements are met. The network is successively partitioned and latencies verified until the required latencies are achieved.

In Chap. 8 we describe measures and metrics for performance monitoring and debugging, and then we describe how to add performance monitoring and debugging to a system. Starting from specified operation timing requirements, a performance monitoring tool should verify latencies for the various paths and operations, operation timing delays, and compliance with prespecified bounds, message, and packet losses. It should also compute statistical information on timing compliance. The chapter ends with the definition of the modules and user interfaces that should be added for performance monitoring and debugging.

Chapter 9 presents experimental results to validate the planning approach, based on an application test bed. A heterogeneous DCS was built by adding wireless sensor networks to a cabled setup. We experimented with various operations and operation timing requirements, planning for network layouts that meet operation timing requirements. Maximum operation timing bounds are specified for sensing,

Fig. 1.1 Types of networks

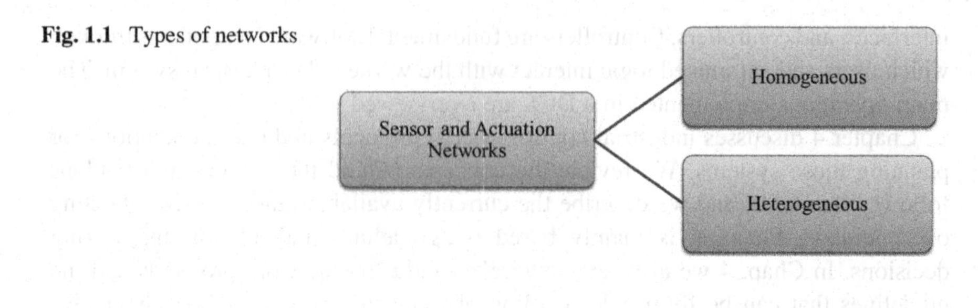

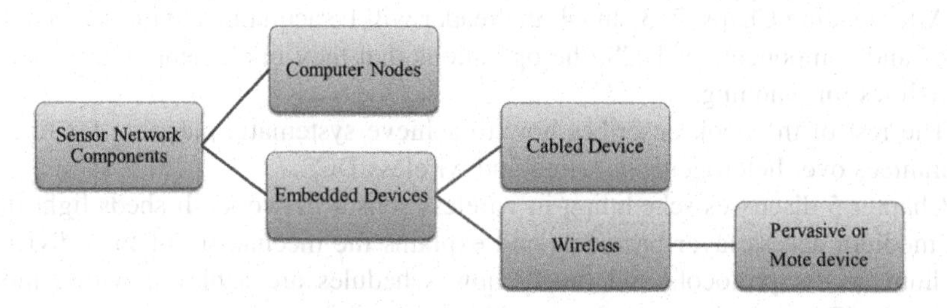

Fig. 1.2 Sensor network components

actuations, and closed loops, including examples of multiple simultaneous closed loops. We show that the models predict latencies adequately and generate useful network layouts. Finally, we use a performance monitoring approach, as described in Chap. 8, to control timing bounds and detect link malfunctions.

After reading this book, the reader will know the main concepts and mechanisms behind DCS as well as the concepts and mechanisms behind wireless sensor networks and scheduling in those types of networks. He or she will also learn how to plan for operation timings adequately in the most difficult scenario of heterogeneous DCS with wired and wireless parts.

Before discussing industrial control systems in more detail in the next chapters, we introduce some useful generic concepts and terms. This book revolves around sensor and actuator networks. These networks can be built with one single platform (hardware and software) or can include different types of platforms. So, they can be homogeneous or heterogeneous (Fig. 1.1).

These networks can be found in different application scenarios. They can be found in DCS, typically used in industrial monitoring and control systems to manage industrial processes, or in other application contexts. In any of these application scenarios, the network may include different classes of hardware and software. It may include embedded devices and computer nodes (Fig. 1.2). In addition, the nodes can be connected using cable infrastructures or wireless links. When wireless links are assumed, the system may include pervasive (or mote) devices. These devices have computation, communication, and programming capabilities but are resource-constrained. Heterogeneity is present because of different device types, different operating software, and different networking protocols.

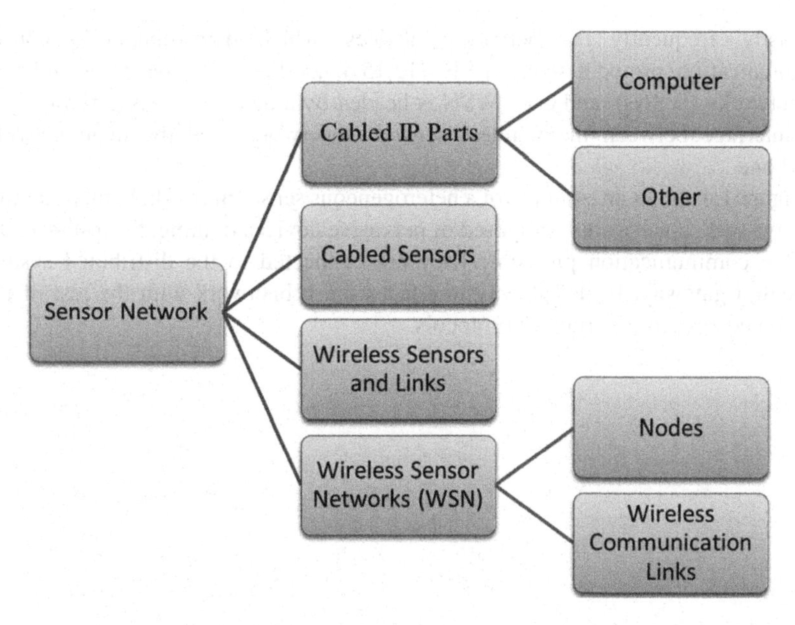

Fig. 1.3 Components of heterogeneous sensor networks

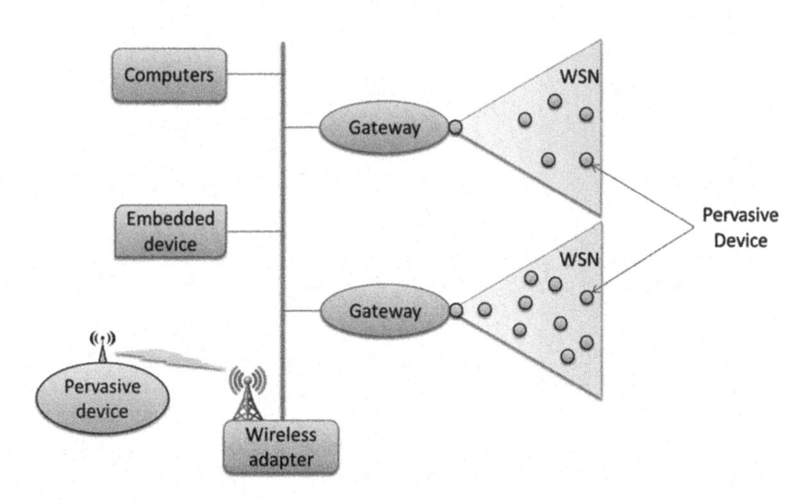

Fig. 1.4 Heterogeneous sensor network

The heterogeneous sensor network organization may include cabled networked parts, computers with diverse functionality, Programmable Logic Controllers, cabled sensors that provide analog signals, wireless sensors and communication links, and wireless sensor networks composed of mote devices and specific communication and routing protocols. Figure 1.3 shows those various components.

Lastly, frequently the pervasive devices, which may operate a non-IP communication protocol (e.g., IEEE 812.15.4, ZigBee), can be organized into subnetworks (WSNs), and each WSN is headed by a gateway. This gateway does the interface between the wireless sensor subnetworks and the main network backbone.

Figure 1.4 shows an example of a heterogeneous sensor network. In this example we have two subnetworks composed of pervasive devices running, for instance, the ZigBee communication protocol. They are connected to the distributed system through a gateway. Each gateway interfaces the subnetwork with the rest of the distributed system that runs IP protocols.

Chapter 2
Industrial Control Systems: Concepts, Components, and Architectures

In this chapter we describe the main concepts, components, and architectures of distributed control systems (DCS). After defining a set of generic concepts and terms related to heterogeneous sensor and actuator networks, we provide an overview of the architectures of industrial networks and DCS. We also review application scenarios for DCS. After those introductory sections, we discuss network topologies, low-level transmission mediums, network communication components, and control components of DCS. The chapter ends with a description of the overall organization of DCS with wireless subnetworks.

2.1 Industrial Networks

Recent advances in industrial networking, such as the incorporation of Ethernet technology, have started to approximate industrial and standard/commercial networks. However, they have different requirements, which results in the need for different architectures. The most essential difference is that industrial networks are connected to physical equipment in some form and are used to control and monitor real-world actions and conditions [1].

Industrial networks generally have a much "deeper" layered layout than commercial networks. The standard/commercial network of a company may consist of branch or office local area networks (LANs) connected by a backbone network or wide area network (WAN), while even small industrial networks tend to have a hierarchy of three or four levels deep.

For example, the connection of instruments to controllers may happen at one level, the interconnection of controllers at the next, and human–machine interfaces (HMI) may be situated above that, with a final network for data collection and external communication sitting at the top. Different protocols and/or physical medium are often used at each level, requiring gateway devices to facilitate communication. Improvements to industrial networking protocols and technology

J. Cecílio and P. Furtado, *Wireless Sensors in Industrial Time-Critical Environments*, Computer Communications and Networks, DOI 10.1007/978-3-319-02889-7_2,

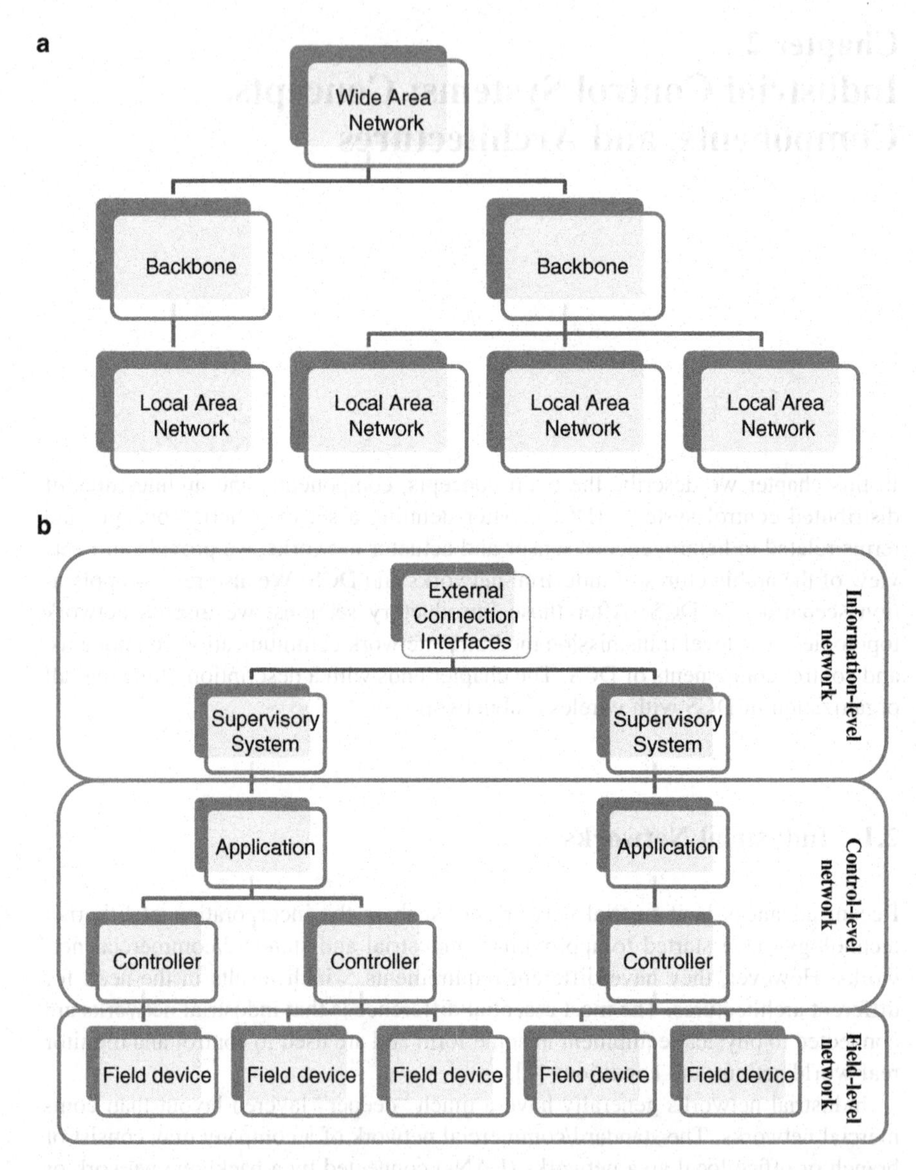

Fig. 2.1 Standard/commercial and industrial network architectures (**a**) standard/commercial network architecture, (**b**) industrial network architecture

have resulted in some flattening of typical industrial hierarchies, especially in the combination of the higher layers.

Figure 2.1a shows an example of a typical standard/commercial network, while Fig. 2.1b shows an example of an industrial network used in DCS.

Typically, industrial subnetworks are classified into several different categories: field-level networks (sensor, actuator, or field device buses), control-level networks (control buses), and information-level networks.

Field-Level Networks – The field-level network is the lowest level of the automation hierarchy. It is used to connect simple, discrete devices with limited intelligence, such as thermo sensors, switch, and solenoid valves to controllers (PLCs or computers). At this level, communication mediums based on parallel, multi-wire cables, and serial interfaces such as the 20 mA current loop have been widely used. In general, these networks connect field devices that work cooperatively in a distributed, time-critical network.

Control-Level Networks – The control-level network is an intermediate level where the information flow consists mainly of the loading of programs, parameters, and data. Typically, control-level networks are used for peer-to-peer connections between controllers such as Programmable Logic Controllers (PLCs), distributed control components, and computer systems used for human–machine interface (HMI).

Information-Level Networks – The information level is the top level of a plant or an industrial automation system. Plant-level controllers gather the management information from the area levels and manage the whole automation system. At the information level, there are large-scale networks, e.g., Ethernet WANs for factory planning and management information exchange. Ethernet networks and gateways can also be used to connect several industrial networks. Typically, information-level networks are used to historical archiving and supervisory control at a high level of supervision.

2.2 Distributed Control Systems Applications

Distributed control systems are employed in many industrial domains including manufacturing, electricity generation and distribution, food and beverage processing, transportation, water distribution, waste water disposal, and chemical refining, including oil and gas. In almost every situation that requires machinery to be monitored and controlled, a distributed control system will be installed. Each industry presents its own set of slightly different but generally similar requirements, which can be broadly grouped into the following domains: discrete manufacturing, process control, building automation, utility distribution, transportation, and embedded systems. They are organized as control architectures, containing a supervisory level of control and overseeing multiple integrated subsystems that are responsible for controlling the details of a process. Process control is achieved by deploying closed-loop control techniques whereby key process conditions are automatically controlled by the system. For automatic control, specific algorithms are applied with proportional, integral, and/or derivative settings. Each algorithm is designed and applied in a manner that should also provide self-correction during process upsets. Those algorithms most frequently run on

Programmable Logic Controllers (PLCs), which are computer-based solid-state devices that control industrial equipment and processes.

The application categories for DCS can be roughly characterized as described next:

Discrete manufacturing assumes that the product being created exists in a stable form between each step of the manufacturing process. An example would be the assembly of automobiles. As such the process can easily be divided into cells, which are generally autonomous and cover a reasonably small physical area. Interconnection of each cell is generally only at a high level, such as at the factory floor controller.

Process control involves systems that are dynamic and interconnected, such as electricity generation. Such systems require interconnection at a lower level and the availability of all plant equipment to function.

Building automation covers many aspects such as security, access control, condition monitoring, surveillance, and heating or cooling. The information being gathered is generally lower, and the distributed control systems are oriented mostly to do supervision and monitoring than control. The large variation in building topology and automation requirements usually results in large variations of architectures.

Utility distribution tends to resemble discrete manufacturing networks in their requirements, despite the fact that the controlled equipment tends to be interconnected. This is mainly because of the large physical distance covered by the distribution system, which makes interconnectivity of the control network more difficult but also increases the time it takes for conditions at one cell to influence another.

Transportation systems are oriented to cover large distances as they deal with the management of trains, monitoring of highways, and the automation of traffic controllers.

Lastly, embedded systems generally involve the control of a single discrete piece of machinery, such as the control networks found in cars. Typically, the distributed control system is designed to cover a very small physical area, but tend to be demanding environments and to have very strict timing and safety requirements.

2.3 Network Topologies

Industrial systems usually consist of hundreds or thousands of devices. As they get larger, network topologies must be considered, featuring subnetworks and main backbones. The most common network topologies used in wired industrial networks are bus, star, ring, or hybrid networks that combine the other ones.

A star configuration, shown in Fig. 2.2, consists of a central node to which all nodes are directly connected. This allows easy connection for small networks, but a hierarchical scheme with multiple subnetworks should be considered once a maximum number of nodes is reached. If a node fails in a star configuration, it does not affect other nodes, unless it is the central node that fails. The star topology may

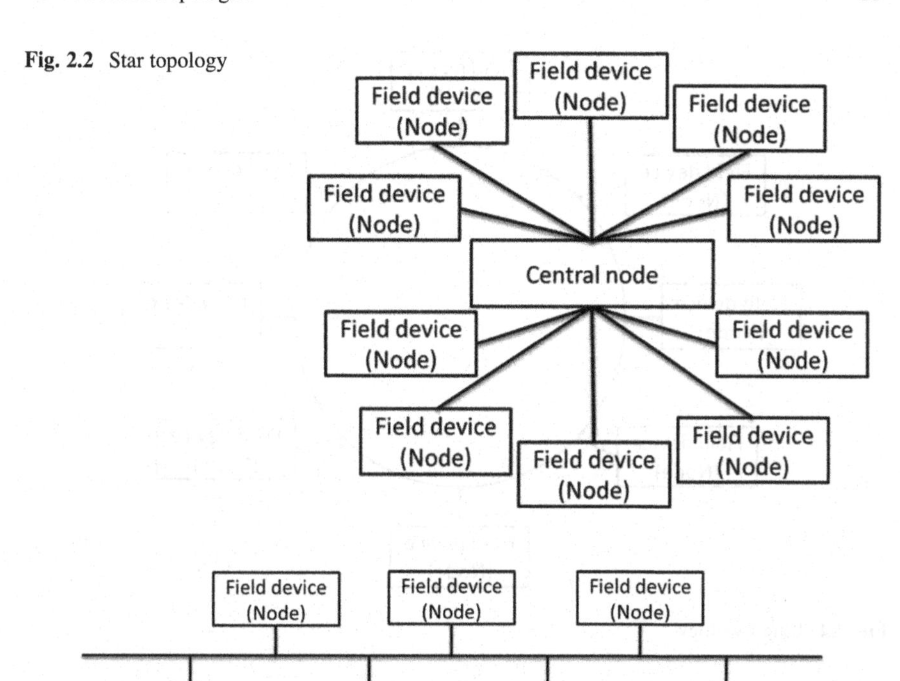

Fig. 2.2 Star topology

Fig. 2.3 Bus topology

have one or more network segments that radiate from the central node. Addition of further nodes is easy and can be done without interrupting network operation.

In the bus topology (Fig. 2.3), each node is directly attached to a common communication channel. Messages transmitted on the bus are received by every node. If a node fails, the rest of the network continues operation, as long as the failed node does not affect the transmission medium.

In the ring topology (Fig. 2.4), a loop is created and the nodes are attached at intervals around the loop. Messages are transmitted around the ring, passing through the nodes attached to it. If a single node fails, the entire network could stop unless a recovery mechanism is implemented.

Most networks used for industrial applications use a hybrid combination of the previous topologies to create larger networks consisting of hundreds, even thousands, of nodes.

Industrial control systems may also include wireless nodes. Those are most frequently organized as star, tree, or mesh networks. The star network topology consists of a single coordinator and leaf nodes. The coordinator is responsible for initiating and maintaining a network. Upon initiation, leaf nodes can only communicate with the coordinator. In the tree topology, there are leaf nodes, relay nodes, and parent nodes. This topology consists of a central node called the root node, and it is the main communications router and interface with other parts of a larger

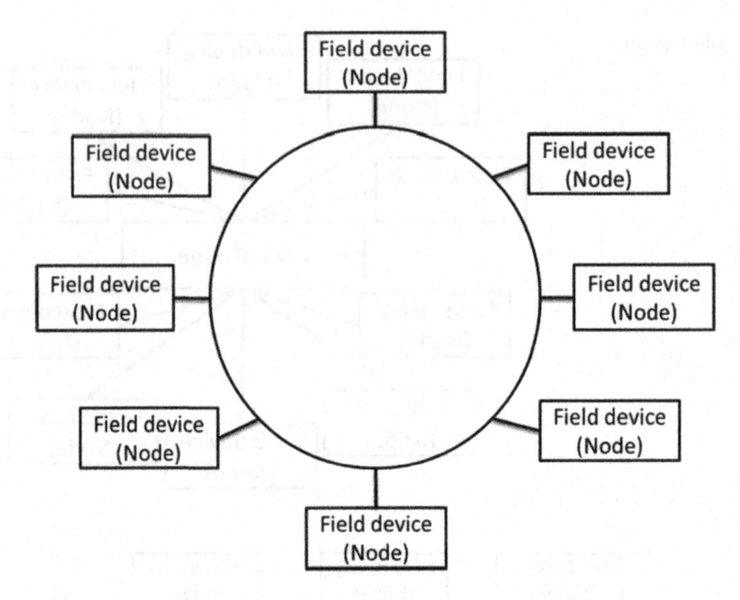

Fig. 2.4 Ring topology

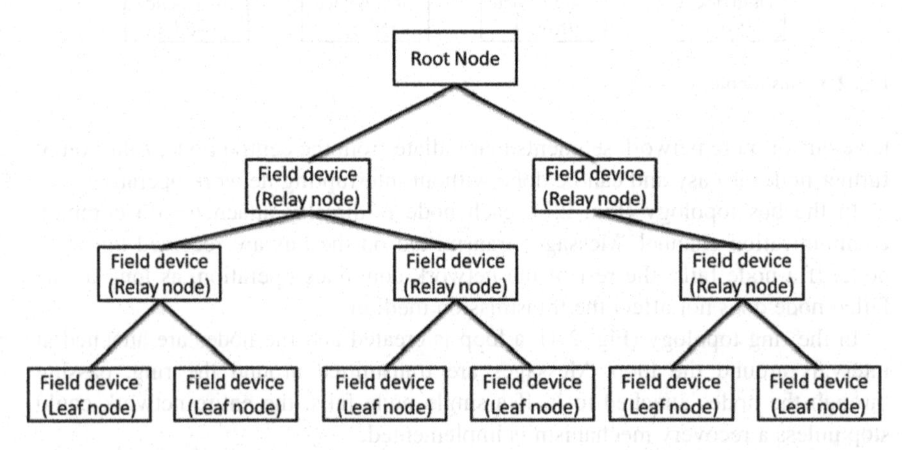

Fig. 2.5 Tree topology

network. One level down from the root node in the hierarchy is a relay node that may have several children nodes connected to it and so on. An example of a tree network is shown in Fig. 2.5.

Lastly, there is the mesh topology. This kind of network topology connects each node to all the nodes within its radio range. It has the potential for reduced power consumption when compared to other topologies, since each node can establish a radio link with the closest neighbors.

2.4 Low-Level Transmission and Mediums

In general, data communication can be analog or digital. Analog data takes continuously changing values, while digital communication data uses only binary values.

Industrial networks require the transmission of both periodically sampled data and aperiodic events such as changes of state or alarm conditions. These signals must be transmitted within a set time period. The sampling period used to collect and transmit data periodically may vary from device to device, according to control requirements, and aperiodic data may occur at any time instant and need to be transmitted to controller elements within specified operation timing bounds. The speed at which processes and equipment operate means that data should be transmitted, processed, and responded to with specific timing requirements, in some cases as close to instantly as possible. Besides real-time requirements, data transmission must also be done in a predictable or deterministic fashion. For a network to be deterministic, it must be possible to predict when a transmission end-to-end will be received, regardless of whether it is a homogeneous or heterogeneous network. This means that the latency of a signal must be bounded. More generically, the end-to-end operation timings, from when some event is triggered to when it is perceived as an alarm or fed into a controller for reaction, must also be bounded.

The transmission methods in industrial networks may be based on baseband, broadband, and carrier band strategies. In baseband transmission a transmission consists of a set of signals that are applied to the transmission medium without frequency modulation. Broadband transmission uses a range of frequencies that can be divided into a number of channels. Carrier band transmission uses only one frequency to transmit and receive information.

Typically, these transmission methods are applied over copper wire, either in the form of coaxial or twisted-pair cable to transmit data over industrial communication networks. Fiber optics and wireless technologies are also being used. Figure 2.6 shows some examples of transmission mediums. In that figure we represent only cabled mediums.

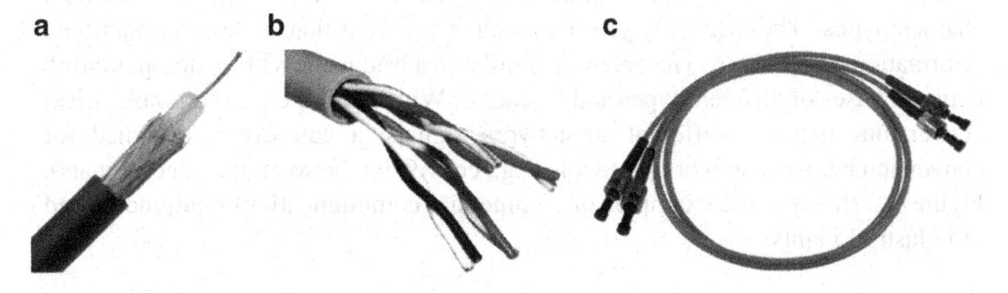

Fig. 2.6 Cabled transmission mediums (**a**) coaxial cable, (**b**) twisted-pair cable, (**c**) fiber-optic cable

Coaxial cables are used for high-speed data transmission over distances of several kilometers. They are relatively inexpensive and can be installed and maintained easily. For those reasons, it is widely used in many industrial communication networks. Twisted-pair cables are used mainly to transmit baseband data at several Mbit/s over short distances. They can be used for high-distance communication, but the transmission speed is decreased according to cable length. Twisted-pair cable has been used for many years and is also widely used in industrial communication networks. It is less expensive than coaxial cable, but it does not provide high transmission throughputs. Fiber-optic cable improves transmission throughput to a few Gbits, and it is free from electromagnetic interference because it is based on light transmission. Digital optical fiber transmission is based on representing the ones and zeros as light pulses. However, the associated equipment required is more expensive, and, if it was used to connect sensors in process plants, separate copper wiring would be required for supplying power to those sensors.

2.5 Network Communication Components

In large industrial and factory networks, a single cable is not enough to connect all nodes together. Industrial communication networks must be divided into several subnetworks to provide isolation and meet performance requirements, mainly timing requirements. The introduction of digital networking simplified the cabling and wirings of the system, leading to further ease of maintenance. Nodes must be able to communicate across several subnetworks within the same industrial network. In this case, additional network equipment is needed to isolate or amplify electrical signals, as well as to make interoperable different communication protocols that can be applied within each subnetwork. Typically, network components such as repeaters, routers bridges, and gateways are used. Next we describe each one.

A repeater, or amplifier, is a device that enhances electrical signals to allow transmission between nodes over greater distances. The router switches communication packets between different network segments, defining the path. Bridges connect between two different subnetworks, which can have different electrical characteristics. These bridges join two subnetworks so that nodes can distribute information across them. The gateway, similar to a bridge, provides interoperability between buses of different types and protocols. While a bridge is used for electrical conversions between different subnetwork signals, a gateway is designed for conversion between network protocols (e.g., conversion between message formats). Figure 2.7 shows some examples of commercial communications equipment used in industrial plants.

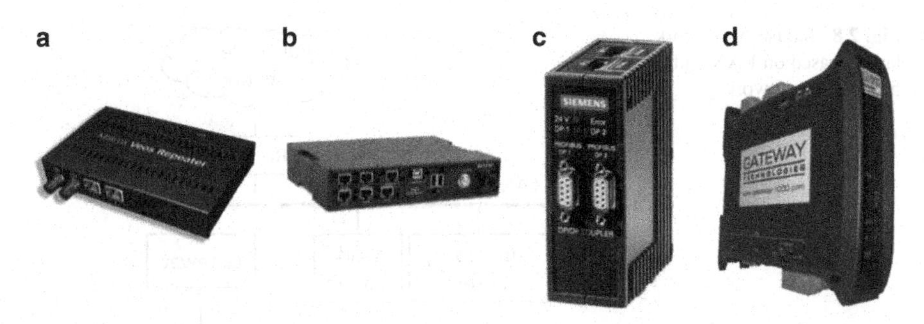

Fig. 2.7 Industrial network communication components (**a**) repeater, (**b**) router, (**c**) bridge, (**d**) gateway

2.6 Control System Components

Figure 2.8 shows an example of an industrial control system layout.

In DCS, the control server hosts the supervisory control software that communicates with lower-level control devices. The control server accesses subordinate control modules over an industrial network. Likewise, the SCADA server is the device that acts as the master in a SCADA system. Remote terminal units and other control devices located at remote field sites usually act as slaves. The remote terminal unit (RTU) is a special-purpose data acquisition and control unit designed to support SCADA remote stations. RTUs are field devices often equipped with wireless radio interfaces to support remote situations where wire-based communications are unavailable.

The Programmable Logic Controller (PLC) is a small industrial computer originally designed to perform the logic functions executed by electrical hardware (relays, switches, and mechanical timer/counters). PLCs have evolved into controllers with the capability of controlling complex processes, and they are used substantially in SCADA systems and DCS. In SCADA environments, PLCs are often used as field devices because they are more economical, versatile, flexible, and configurable than special-purpose RTUs. The actual programming of a PLC is done using specialized programming software and a physical connection to a dedicated programming port on the device. Many PLCs also supports programming through a network connection. Current solutions for programming PLCs are graphic-based (high-level programming language) to allow easy programming.

The fieldbus network links sensors, PLCs, and other devices or controller within the control system. Use of fieldbus technologies eliminates the need for point-to-point wiring between the controller and each device. The devices communicate with the fieldbus controller using a variety of protocols. The messages sent between the sensors and the controller uniquely identify each of the sensors.

Field devices (FD) are "smart" sensor/actuator containing the intelligence required to acquire data, communicate to other devices, and perform local processing and control. A field device can combine an analog input sensor, analog

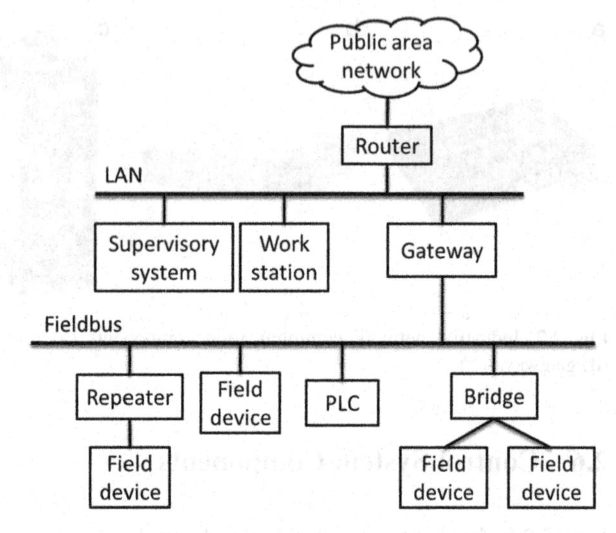

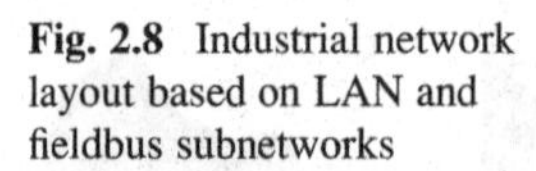
Fig. 2.8 Industrial network layout based on LAN and fieldbus subnetworks

output, communication capabilities, and program and data memory in a single device. The use of FDs in SCADA and DCS systems allows the development of distributed algorithms for automatic process control.

The human–machine interface (HMI) is composed of software and hardware parts that allow human operators to monitor the state of a process under control, modify control settings, and manually override automatic control operations in the case of an emergency. The HMI also allows an operator to configure closed-loop algorithms and parameters in each controller. The HMI also displays process status information, reports, and other information to operators, administrators, or managers. The data repository is a centralized database for logging all process information within an industrial control system. Information stored in this database can be accessed to support various analyses, from statistical process control to process planning. The input/output (IO) server is a control component responsible for collecting, buffering, and providing access to process information from control subcomponents such as PLCs, RTUs, and FDs. An IO server can reside on the control server or on a separate computer platform. IO servers are also used for interfacing third-party control components, such as an HMI and a control server.

2.7 Overall System Organization

An industrial control system is typically made of various subsystems. Figure 2.9 shows an example. It includes resource-constrained sensor nodes and more powerful nodes such as PLCs, gateways, or computer nodes. A heterogeneous industrial control system with wired and wireless components can include wireless sensor nodes organized into wireless sensor subnetworks, as shown in the figure.

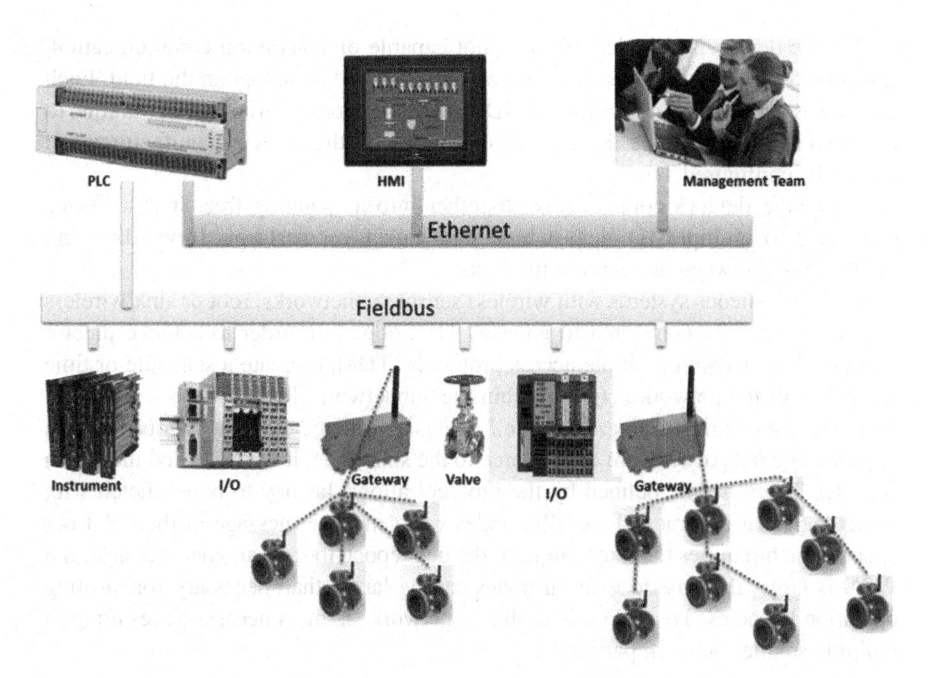

Fig. 2.9 Organization of distributed control system with heterogeneity

Fieldbus networks used in industrial control systems are sometimes denoted as H1 and HSE. The first, H1, is a digital, serial, two-way communication system running at 31.25 kbit/s, which interconnects field equipment such as sensors, actuators, and controllers. The second protocol, high-speed Ethernet (HSE), uses 10 or 100 Mbps Ethernet as the physical layer and provides a high-speed backbone for the network. Its large capacity to move data, along with the inherent fieldbus functionality, and publish/subscribe access, fits in with plant-wide integration in process industries.

The fieldbus uses a token ring protocol to control communication with devices. It prevents data communication collisions between two nodes that want to send messages at the same time, using a special three-byte frame called a token that travels around the network (ring). Token possession grants the possessor permission to transmit on the medium. Token ring frames travel completely around the loop.

There are three types of devices on a fieldbus H1 network: link masters, basic devices, and H1 bridges.

A link master device is capable of controlling the communications traffic on a link by scheduling the communication on the network. Every fieldbus network needs at least one link master-capable device. However, only one link master will actively control the bus at a given time. The link master that is currently controlling the bus is called the Link Active Scheduler (LAS). If the current Link Active Scheduler fails, the next link master will take over transparently and begin controlling bus communications where the previous Link Active Scheduler left off. Thus, no special configuration is required to implement redundancy.

A basic device is a device which is not capable of scheduling communication. It is a field device that, typically, connects sensors and actuators on the field. Each device has a unique identifier (device ID), which includes a serial number unique to the device. Typically, the device ID is assigned by the device manufacturer and cannot be configured.

H1 bridge devices connect links together into a spanning tree. It is a device connected to multiple H1 links whose data link layer performs forwarding and republishing between and among the links.

In heterogeneous systems with wireless sensor subnetworks, root or sink wireless nodes are typically connected to gateway H1 bridges. In order to achieve precise timings, time-division multiple access protocols (TDMA) create a schedule or time frame for all the network activity within the subnetwork: Each node is assigned at least one slot in a time frame. The time frame is considered to be the number of slots required to get a packet from each source to the sink node. It is also called the epoch size (E). The schedule defined by the protocol allows latency to be predicted with some degree of accuracy. Typically, nodes will send one message in their slot per epoch, which requires them to wait until the next epoch to send another message. If a WSN is large, the time to visit all nodes can be larger than necessary for meeting operation latencies. To avoid this problem, network sizing is necessary, resulting in multiple smaller network partitions.

Reference

1. Stouffer K, Scarfone K (2011) Guide to industrial control systems (ICS) security recommendations of the national institute of standards and technology. Nist Special Publication, vol 800, No. June, pp 1–156

Chapter 3
Distributed Control System Operations

An industrial distributed control system is also a programmable electronic cyber-physical system prepared to perform a set of operations continuously. It should implement real-time monitoring, continuous recording and logging of plant status, and process parameters. It should also provide information to operators regarding the plant status and process parameters and allow operator control over the processes, including the capability to change the plant status. The DCS should implement automatic process control and batch/sequence control during start-up, normal operation, shutdown, and disturbance, i.e., control within normal operating limits. It should also be able to detect hazardous conditions and to terminate or mitigate hazards (i.e., control within safe operating limits). Finally, it should enforce automatic or manual hazard prevention, by means of control actions over conditions that might initiate the hazard. In this chapter we define the operations that the system should implement and the main elements involved.

Figure 3.1 shows an operation taxonomy that can be used to classify underlying operations of the DCS, over which all higher-level operations should be defined.

Configuration involves commanding a node in the DCS, to configure it or to change its status. The node can be a workstation, a control station, a programmable logic controller, or a smart device (sensor or actuator). Examples of configurations can be to modify a set point or a sampling rate or to modify what is displayed.

Monitoring is typically implemented by means of some sensors sensing physical variables periodically (sampling rate) and sending the data measures to a control station, where they will be processed and delivered to users in some processed, user-friendly manner. The monitoring operation can also be event-based. In this case the operation executes only when an external event occurs. Alarms are a specific kind of operation, alerting operators to plant conditions, such as deviation from normal operating limits and to abnormal events requiring timely action or assessment.

J. Cecílio and P. Furtado, *Wireless Sensors in Industrial Time-Critical Environments*, Computer Communications and Networks, DOI 10.1007/978-3-319-02889-7_3,

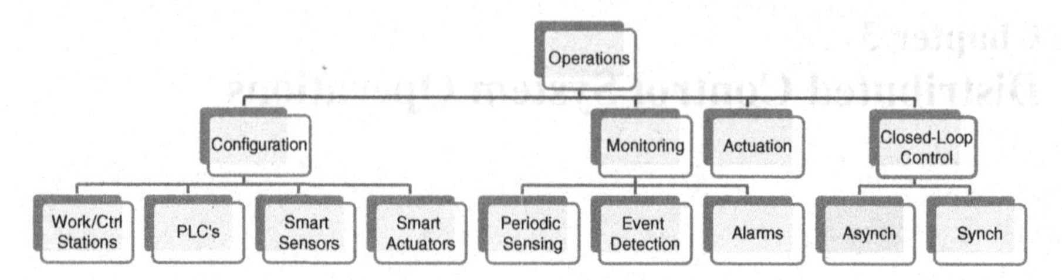

Fig. 3.1 Underlying basic distributed control operations

Actuations are the operations that allow a plant operator to directly modify the status of some physical subsystem, typically by sending commands to a node to actuate over a DAC connected to some physical process.

System and process alarms are essential to safe and efficient operation in a DCS. Alarms should be specifically designed to attract operator attention during operation upsets and to allow operators to have a clear view of unit operating condition. Alarms should be presented either as part of alarm lists or embedded in operator displays, using graphic illustrations. In addition to the displays, sounds can be used to alert operators in case of emergency.

There are some issues that should be taken into consideration concerning alarms. Firstly, it should be possible to configure alarm settings, and both required and current alarm settings should be well documented. Alarm system management procedures should be available, including alarm description and parameters, as well as alarm response procedures. Secondly, alarm operation should ensure that repeating alarms do not result in operator overload. Applicable alarm processing techniques include grouping and first-up alarms, eclipsing of lower-grade alarms when higher-level ones activate, suppression of selected alarms during certain operating modes, automatic alarm load shedding, and shelving. However, care is necessary in the use of shelving or suppression, to ensure that alarms are returned to an active state when they are relevant to plant operation. Thirdly, the human interface for alarms should be suitable and adapted to intended use. Alarms may be presented either on annunciator panels, individual indicators, visual display unit (VDU) screens, or programmable display devices. Alarm lists should be carefully designed to ensure that high-priority alarms are readily identified, that low priority alarms are not overlooked, and that the list remains readable even during times of high alarm activity or with repeated alarms. Alarms should be prioritized in terms of which alarms require the most urgent operator attention and presented within the operator's field of view. Finally, each alarm should provide sufficient operator information for the alarm condition, plant affected, action required, alarm priority, time of alarm, and alarm status to be readily identified.

Closed-loop control operations involve sensing, sending data measures to supervision control logic, processing them, determining actuation commands to send to an actuator, sending the actuation commands, and actuating. Similar to monitoring, supervision control logic can react based on events (asynchronous control) or with a predefined periodicity (synchronous control). Asynchronous control can be defined as "upon reception of a data message, the supervision control logic computes a command and sends it to an actuator." The control computations are based on events. For instance, we can configure a node to send data messages only if a certain threshold condition is met. Synchronous control can be defined as a periodic control, where the periodicity of computations is defined by the requirements of more or less complex controller logic. The supervision control logic runs in specific instants (period). Typically, this type of controller involves multiple samples for the computation of the actuation commands, and multiple sensors may participate in sensing. Sometimes control loops are nested, whereby the set point for one loop is based on the process variable determined by another loop. Supervisory-level loops and lower-level loops operate continuously over the duration of a process with cycle times ranging from the order of milliseconds to minutes.

3.1 System Components for Distributed Control Operation

The distributed control system runs on a specific environment and has a set of components that include human and plant interfaces, logic elements, and management systems. The environment is the physical place in which the control system (including the operator) is required to work, including environmental conditions (humidity, temperature, pressure, flammable atmospheres) and external influences such as electromagnetic radiation and hazards, which might affect the operation of the control system. The human interface is the interface with operators. It may comprise a number of input and output components, such as controls, keyboard, mouse, indicators, enunciators, graphic terminals, mimics, audible alarms, and charts. The plant interface comprises inputs (sensors), outputs (actuators), and communications (wiring, fiber optic, analog/digital signals, fieldbus, and/or wireless links). The logic elements are all elements that implement the control operations. They may be distributed and linked by communication links. A whole distributed control system (DCS) can be called a logic element, as can also individual components, such as relays, discrete controllers, computers, or programmable logic controllers (PLC). Logic elements are designed to perform continuous control or discrete state changes (e.g., start-up/shutdown).

The basic elements for operation of an industrial control system are shown in Fig. 3.2 [1]. The diagram includes human–machine interfaces (HMI), remote diagnostics and maintenance tools, controllers, sensors, and actuators together with the main processes that we want to control.

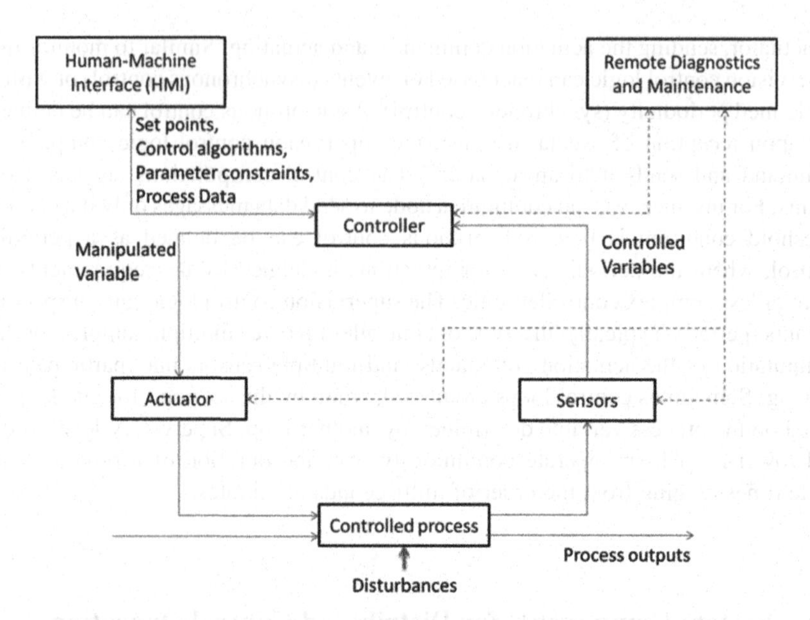

Fig. 3.2 Operations diagram

3.2 Human–Machine Interfaces, Diagnostics, and Maintenance

Operators and engineers monitor and configure set points, control algorithms, and parameters in controllers using human–machine interfaces (HMI). The HMI also displays process status and historical information. A typical set of graphical functions includes single point display for monitoring and control, real-time and historical trend displays, alarm status and history displays, and, finally, criteria-based threshold lists. Criteria such as signal type, controller assignment, sequence of events (SOE), and alarm status can be used as filtering criteria in those systems.

Custom-built dynamic displays are frequently designed, based on user/plant specifications and requirements.

Besides HMIs, industrial control systems also include diagnostics and maintenance utilities to prevent, identify, and recover from abnormal operation or failures. Typically, the diagnostics and maintenance components provide an interface for plant operators, engineers, and maintenance personnel to gather critical information when and where it is needed. This includes real-time and historical process data, intuitive alarm indicators, sequence of events, and reports. The information shown is oriented toward fault diagnosis and planning maintenance tasks.

3.3 Sensors and Actuators

Sensors and actuators are direct interfaces with the physical systems that they control. A sensor is defined as a device that converts a physical measurement into a readable output. By extracting information from the physical world, sensors provide inputs to industrial control systems. The role of the sensor in an industrial control system is therefore to detect and measure some physical variable and to provide information to the control system regarding a related property of the system under control.

Nowadays, many industrial control systems include smart sensors. The term "smart sensor" implies that some degree of signal conditioning is included in the same package as the sensor. Typically, those sensors are more sophisticated devices, able to do signal processing, displaying and diagnosing, and exhibiting self-calibration features.

There are many ways to classify sensors. Classifications can be found in the literature based on energy, transduction mechanism (e.g., thermocouple or piezo-electric materials), and modulation mechanisms (e.g., thermistor or piezoresistor). Table 3.1 shows some examples of sensors.

The classification of those sensors can be done according to which physical variable the sensor is able to measure. For instance, Table 3.2 shows a classification of sensors according to their sensing principle.

Sensors must go through a process of calibration, testing, and quality measurement before certification and continued use. In the case of smart sensors, they also require adequate diagnostic coverage and fault tolerance, as well as measures to protect against systematic failures. The following characteristics should be tested and taken into account concerning sensors:

Table 3.1 Example of sensors

Sensor	Sensed quantity
Thermometer	Temperature
Psychrometer	Humidity
Barometer	Pressure
Photodiode	Light
Odometer	Distance
Laser	Distance

Table 3.2 Classification of sensors based on their sensing principle

Sensor	Sensed quantity
Thermal	Thermometer, thermocouple, thermistor, air flow sensors
Mechanical motion	Quartz clock, spring balance, odometer, piezoresistive pressure sensor, accelerometer, gyro
Optical energy	Photodiode, CCD camera, color sensor
Magnetic field	Compass, Hall effect, inductive proximity sensor
Electric field	Electrostatic voltmeter, field-effect transistor

- Cross sensitivities to other fluids which might be present in a sensed process.
- Reliability of sampling systems.
- Signal conditioning (e.g., filtering), which may affect the sensor response times.
- Degradation properties of measurement signals (distance between sensor and transmitter may be important).
- Accuracy, repeatability, hysteresis, and common mode effects (e.g., effects of gauge pressure or temperature on differential pressure measurement).
- Integrity of process connections and sensors for containment.
- Protection against systematic failures on programmable sensors/analyzers. The sensors capabilities regarding protection against systematic failures should be tested. The measures taken will depend on the level of variability and track record of the software. Sensors with limited variability of the software, which are extensively proven in use, may require no additional measures other than those related to control of operation, maintenance, and modification.

Actuators are the output interface of the control system with physical systems that it controls. They are most frequently electrical equipments that modify the state of physical systems. In industrial control, controllers output commands to actuators to apply corrections from measured deviations from the set point. Examples of actuators include motors, valves (control and isolation), positioners, and switched power supplies. Actuators may also include programmable control elements, particularly within positioners; variable speed drives; and motor control centers. Modern motor control centers may also use programmable digital addressing, adding significant flexibility.

3.4 Process Monitoring and Controllers

Controllers provide capabilities to monitor and control plant processes. Typically, controllers are designed once using specific control algorithms tailored for the process that needs to be controlled. The controller parameters are then configured and tuned by engineers, using engineering station user interfaces. Nowadays the control software supports powerful control applications through straightforward configuration of function blocks. A wide range of process control capabilities includes continuous control, logic control, sequence control, and advanced control. Both online and off-line configuration is supported.

Controllers may compare a set of data to set points, perform simple addition or subtraction functions to make comparisons, or perform other complex mathematical functions. Controllers always have an ability to receive input signals, perform mathematical functions with the inputs, and produce output signals.

There are a number of important concepts related to process control. A process variable is a condition of the process fluid (a liquid or gas) that can change the manufacturing process in some way. Common process variables include pressure, flow, level, temperature, and light density. The set point is a value for a process

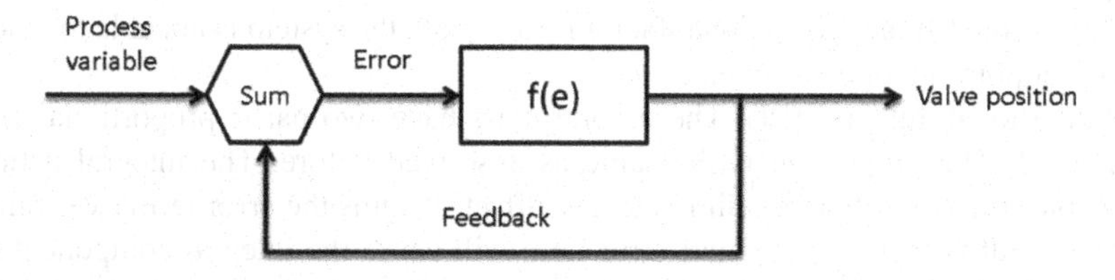

Fig. 3.3 Controller algorithm example

variable that is desired to be maintained. For example, if a process temperature needs to be kept within 5 °C of 70 °C, then the set point is 70 °C. A temperature sensor can be used to help maintain the temperature at set point. The sensor is inserted into the process, and a controller compares the temperature readings from the sensor to the set point. If the temperature reading is 80 °C, then the controller determines that the process is above set point and signals, for example, a fuel valve of the burner to close slightly until the process cools to 70 °C. Set points can also be maximum or minimum values.

Error is the difference between the measured variable and the set point and can be either positive or negative. The objective of any control scheme is to minimize or eliminate error. Therefore, it is imperative that error be well understood. Any error can be seen as having two major components: Magnitude represents the deviation between the values of the set point and the process variable. The magnitude of error at any point in time compared to the previous error provides the basis for determining the change in error. The change in error may be an important input, depending on the process control; duration refers to the length of time that an error condition exists. The offset is a sustained deviation of the process variable from the set point.

A control algorithm is a mathematical expression of a control function. Using the temperature example in the definition of set point, V represents the fuel valve position, and e is the error. The relationship in a control algorithm can be expressed as

$$V = f(\pm e)$$

The fuel valve position (V) is a function (f) of the error. Figure 3.3 shows an example of a simple controller algorithm.

Control algorithms can be used to calculate the requirements of much more complex control loops than the ones described here. In more complex control loops, questions such as "how far should the valve be opened or closed in response to a given change in set point?" and "how long should the valve be held in the new position after the process variable moves back toward the set point?" need to be answered. There are three main types of control algorithms in the literature:

- Proportional (P). In the proportional control algorithm, the controller output is proportional to the error signal, which is the difference between the set point and the process variable. In other words, the output of a proportional controller is the product of the error signal and the proportional gain. Increasing the proportional factor will increase the response time of the controller. However, if the proportional factor is too high, the system will become unstable and oscillate around the set point. If the proportional factor is too small, the system is not able to reach the set point in a given time.

- Proportional–Integral (PI). These controllers have two parts: proportional and integral. The proportion is the same as described before. The integral action enables the controllers to eliminate any offset. It sums the error term over time. The result is that even a small error term will cause the integral component to increase slowly. This means there is a continually increase over time until the error term reaches zero.
- Proportional–Integral–Derivative (PID). These controllers have the previous functions and add a derivative term. The derivative term predicts system behavior and thus improves settling time and stability of the system. However, if the noise influence is high, the derivative action will be erratic and will degrade control performance. By using all proportional, integral, and derivative parts together, process operators can achieve rapid response to major disturbances with derivative control, hold the process near the set point without major fluctuations of proportional control, and eliminate offsets, a typical capability of integral control.

Not every process requires a full PID control strategy. If a small offset has no impact on the process, then proportional control alone may be sufficient. PI control is used where no offset can be tolerated, noise (temporary reading spikes that do not reflect the true process variable condition) may be present, and where excessive time (time before control action takes place) is not a problem.

A closed-loop control exists where a process variable is measured, compared to a set point, and action is taken to correct any deviation from set point. An open-loop control exists where the process variable is not compared, and action is taken not in response to feedback on the condition of the process variable, but is instead taken without regard to process variable conditions. For example, a water valve may be opened to add cooling water to a process to prevent the process fluid from getting too hot, based on a preset time interval, regardless of the actual temperature of the process fluid.

Before process automation became a reality, people, rather than machines, performed many of the process control tasks. For example, a human operator can keep watch of a level gauge and close a valve when the level reaches the set point. This can be done manually, by physically moving some actuator part or more commonly by commanding the operation through the industrial control system, using some user interface. Control operations that involve human action to make an adjustment are called manual control systems. Conversely, control operations in which no human intervention is required, such as an automatic valve actuator that responds to a level controller, are called automatic control systems. Manual control is typically present even in totally automated systems, for instance, for instant shutdown in case of major malfunctions and other emergency situations.

Reference

1. Kalani G (2002) Industrial process control: advances and applications. Butterworth-Heinemann, London, p 182

Chapter 4
Industrial Protocols and Planning Considerations

In this chapter we review briefly industrial control system protocols and discuss considerations concerning planning within those protocols. Some of the main concepts behind the fieldbus and fieldbus follower standards are reviewed, as well as guidelines for planning. We also review wireless industrial network protocols and the guidelines that can be followed regarding planning within those networks.

4.1 The Fieldbus Standard

The fieldbus [1–3] defines an open international digital communication standard for industrial automation [4]. It was proposed to avoid the limitations of proprietary protocols, including slow transmission speeds and differing data formats. Also related to the fieldbus, the Fieldbus Foundation is an international organization that was established in 1994 to promote a single international fieldbus to both users and vendors, to create a foundation fieldbus specification, to provide technologies for fieldbus implementation, and to install an infrastructure to achieve interoperability. The Foundation Fieldbus standard is a subset of the IEC/ISA standard (IEC61158 and ISA s50.02).

The Foundation Fieldbus defines "Fieldbus as a digital, two-way, multi-drop communication link among intelligent measurement and control devices." It was designed to reduce the life cycle cost of production line and then total cost of the plant operation and maintenance. It integrates plant assets on a single plant automation system with digital communication networks. Also, devices from multiple suppliers can be connected using the fieldbus without any additional custom software.

The fieldbus offers reduced installation and material cost, by replacing the traditional one-to-one wiring scheme with networking or multi-drop configurations, while intelligent field instruments make commissioning and plant start-up much faster and less expensive. It also defines two main communication network layers

J. Cecílio and P. Furtado, *Wireless Sensors in Industrial Time-Critical Environments*, Computer Communications and Networks, DOI 10.1007/978-3-319-02889-7_4,

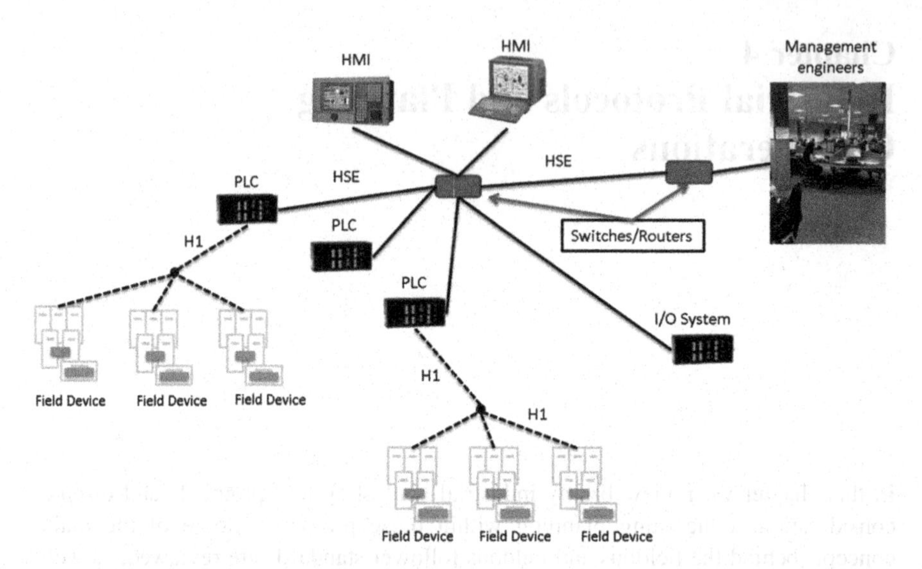

Fig. 4.1 Industrial network and components

(H1 and HSE). Figure 4.1 shows an example of an industrial control network, with some of its components and those layers. The first layer, H1, is an intrinsically safe layer supporting transmissions of up to 31.25 kbit/s. At this level, Foundation Fieldbus networks are based on network segments (up to 65,000 possible segments, where each segment may have distances up to 1,900 m and 240 nodes in it). In the figure, multiple field devices are connected to Programmable Logic Controllers (PLCs) in this level.

The second layer HSE (high-speed Ethernet) allows transmissions up to 100 Mbit/s and was designed to connect human–machine interfaces (HMIs), systems management engineers, and management processes to the main control network. It defines message formats for client/server, publisher/subscriber, event notification applications, and HMIs.

In the next subsections we briefly review the communication stack, fieldbus device definition, and system management protocol within the fieldbus. More detailed information on fieldbus details can be obtained from the literature and other resources [1–3].

4.1.1 Communication Stack

Fieldbus communication stack is specified according to the simplified OSI model, consisting of three layers: physical layer, data link layer, and application layer.

Physical Layer – The physical layer defines the mechanisms to transmit and receive electric or optic signals to/from a medium. It concerns wiring, signals, waveforms,

Table 4.1 Spur length (Rule 4)

Devices on the fieldbus	Spur length (m)
1–12	120
13–14	90
15–18	60
19–24	30
25–32	1

voltages, and other details related to electricity and optics mechanisms. Though the IEC/ISA SP50 standard [5] specifies various medium with various speeds, the fieldbus chose its own subset that is adequate for the needs of industrial automation applications (low speed wire, fiber medium, or Ethernet). A 31.25 kbit/s physical layer is the most common since IEC and ISA approved it in 1992. Its devices can be powered directly from the fieldbus and can operate on wiring that was previously used for 4–20 mA devices. A minimum amplitude and worst waveform of a received signal at a device at any place of the fieldbus network is also specified by the IEC/ISA standard. The physical layer receiver circuit must be able to receive this signal. The ISA SP50 committee also created a set of wiring rules to simplify the network design and guarantee signal quality at any receiver.

Rule 1: The number of devices on a fieldbus is between 2 and 32.

Rule 2: The cable is a twisted pair of individual shield (type A) with 18 AWG wires.

Rule 3: Overall cable length must not exceed 1,900 m.

Rule 4: The length of the cable of each fieldbus device connected to a segment (Spur) cannot exceed 120 m. The maximum number of devices decreases according to the Spur length, as shown in Table 4.1.

Rule 5: When you use multiple twisted pairs with overall shield (type B) with 22 AWG wires, the total length decreases to 1,200 m.

Other cable types are also defined but not recommended [3].

Intrinsic safe installation is important for plants where explosive gases exist. It is the rule to design and install devices in hazardous areas, in order to prevent an explosive gas being ignited by electric discharge or the surface temperature of a device. An intrinsic safe device must be carefully designed to prevent ignition even when a single failure of its component takes place. It must include a barrier installed to separate the hazardous area from the safe area. A barrier strictly limits the voltage, current, and power fed to a device installed in the hazardous area. Therefore, a device must be operational with the restricted power supply. Devices and barriers must meet the same design criteria provided by safety organizations (IEC, FM, CENELEC, PTB, etc.).

At upper level of the fieldbus, gateways are used to interconnect 31.25 kbit/s fieldbuses and make them accessible to a high-speed Ethernet (HSE) backbone running at 100 Mbit/s. Since the HSE is based on standard Ethernet protocols (e.g., TCP/IP, SNTP, SNMP), commercial off-the-shelf HSE equipment such as switches and routers is used to create larger networks. Moreover, all parts of the HSE network can be made redundant to achieve the fault tolerance level needed by the application.

Table 4.2 Maximum data size for each priority

Priority	Maximum data size
URGENT	64 bytes
NORMAL	128 bytes
TIME_AVAILABLE	256 bytes

Data Link Layer – The data link layer defines mechanisms to transfer data from a node to the other nodes. It manages the priority and order of data requests.

The most important functionality of the data link layer is the medium access control (MAC) of the fieldbus. Since all devices on the same cable receive the same physical layer signal, only one of them is allowed to transmit a signal at a time.

The Link Active Scheduler (LAS) has the role to control medium access. The right to send data is called a "token." The LAS possesses the token and gives it to another device to allow it to send messages. The token is then returned to the LAS for further medium access control. The LAS is responsible for scheduled communication, which is necessary to link function blocks. Function blocks are distributed applications operating in a synchronized manner. The LAS manages the communication part of the synchronized data transfer. A function block output parameter is a "publisher" of data, and other function blocks that receive this data are called "subscribers." The LAS controls periodic data transfer from a publisher to subscribers using the network schedule.

Since application messages have various levels of urgency, the data link layer also supports a mechanism to transmit messages according to their urgency. There are three levels of "priority": URGENT, NORMAL, and TIME_AVAILABLE. An URGENT message is transmitted immediately, even when other messages of NORMAL or TIME_AVAILABLE priority are in the waiting queue.

According to IEC and ISA SP50 standard, Table 4.2 shows the maximum data size allowed for each priority.

Another important aspect related with data link layer is the addressing scheme. Every fieldbus device must have a unique network address and physical device tag for the fieldbus to operate properly. To avoid the need for address switches on the instruments, the LAS can perform assignment of network addresses automatically. The sequence for assigning a network address to a new device is as follows:

- A physical device tag is assigned to a new device via a configuration device (off-line).
- Using default network addresses, the LAS asks the device for its physical device tag. The physical device tag is used to look up the new network address in a configuration table, and then, the LAS sends a special "set address" message to the device which forces the device to move to the new network address.
- The sequence is repeated for all devices that enter the network at a default address.

Application Layer – The application layer consists of two sublayers: the Fieldbus Access Sublayer (FAS) manages data transfer, and Fieldbus Message Specification (FMS) encodes and decodes user data. The Fieldbus Access Sublayer (FAS) implements secure communication. It provides three communication models for applications: client–server, publisher–subscriber, and source–sink. In the client–server model, an application called “client” requests another application called “server” to do a specific action through FMS. When the server finishes the requested action, its result is transferred to the client. It is a one-to-one, two-way communication. The publisher–subscriber model is designed to link function blocks. When a publishing function block runs, its output data is stored in the buffer of the publisher. Then the LAS sends a request to the publisher to force it to transfer the data. Subscribers receive this data and give the data to the subscribing function blocks. The source–sink model is designed to broadcast messages. It is a one-to-many, one-way communication without schedule. This model is sometimes called “report distribution model.” A source transfers a message in the queue to an assigned global address (address of the sink) when the device has the token. A sink can receive messages from many sources if they are configured to send messages to the same global address.

4.1.2 Fieldbus Device Definition

The Fieldbus Foundation has defined a standard user application based on “blocks.” Blocks are representations of different types of application functions and operations.

Function blocks (FB) provide the control system behavior. The input and output parameters of function blocks can be linked over the fieldbus. The execution of each function block is precisely scheduled. There can be many function blocks in a single user application. Function blocks can be built into fieldbus devices as needed to achieve the desired device functionality. For example, a simple temperature transmitter may contain an Analog Input function block. A control valve might contain a PID function block as well as the expected Analog Output block. The exact function of a fieldbus device is determined by the arrangement and interconnection of blocks.

The device functions are made visible to the fieldbus communication system through the user application virtual field device (VFD). This VFD provides access to the Network Management Information Base (NMIB) and to the System Management Information Base (SMIB). NMIB data includes Virtual Communication Relationships (VCR), dynamic variables, statistics, and Link Active Scheduler (LAS) schedules (if the device is a link master). SMIB data includes device tag and address information and schedules for function block execution. The VFD object descriptions and their associated data are accessed remotely over the fieldbus network using Virtual Communication Relationships (VCRs).

4.1.3 System Management Protocol

Operations inside a fieldbus system must execute at precisely defined intervals and in the proper sequence for correct control of the system. System management functionality synchronizes execution of operations and communications of nodes' data and parameters on the fieldbus. It also handles other important system features, such as publication of the time of day to all devices, including automatic switchover to a redundant time publisher, automatic assignment of device addresses, and searching for parameter names or "tags" on the fieldbus.

All of the configuration information needed by system management, such as the operation schedule, is described by object descriptions in the network and system management virtual field device (VFD) in each device. This VFD provides access to the System Management Information Base (SMIB) and also to the Network Management Information Base (NMIB).

4.1.4 Time Scheduling and Clocks

A schedule building tool is used to generate function block and Link Active Scheduler (LAS) schedules. The schedules contain the start time offset from the beginning of the "absolute link schedule start time." All devices on the fieldbus know the absolute link schedule start time. During function block execution, the LAS sends a token message to all devices so that they can transmit their unscheduled messages such as alarm notifications or operator set point changes.

The fieldbus also supports an application clock distribution function. The application clock is usually set equal to the local time of day or to Universal Coordinated Time. System management has a time publisher, which periodically sends an application clock synchronization message to all fieldbus devices. The data link scheduling time is sampled and sent with the application clock message, so that the receiving devices can adjust their local application times. Between synchronization messages, application clock time is independently maintained in each device based on its own internal clock.

Application clock synchronization allows the devices to timestamp data throughout the fieldbus network. If there are backup application clock publishers on the fieldbus, a backup publisher will become active if the currently active time publisher fails.

4.1.5 Planning for Fieldbus Networks: Some Guidelines and Previous Work

Carefully planned segments are the basis for reliable operation in the fieldbus. When planning a project, it is necessary to determine how many devices can be allocated to each fieldbus segment. According to rule 4 of IEC/ISA, the length of each fieldbus device wire connected to a segment cannot exceed 120 m, and the maximum number of devices is also limited, as defined previously in Table 4.1. The physical layer also restricts the number of devices. The data link layer specifies the address to be 8 bits in length. Because some of these are reserved for special purposes, the actual address range for field devices is 232.

The bus cycle time of operation is an important ingredient in planning a fieldbus segment because it influences the overall system responsiveness. It is determined from the following points:

- Fieldbus system used (i.e., Foundation Fieldbus H1 or PROFIBUS PA)
- DCS system used
- Number of devices on the segment
- Amount of data to be transmitted
- Time required for acyclic data exchange (e.g., configuration and diagnostic data)

Applying real-time protocols and planning methodologies, latencies, and delays can be assessed (determined or, at least, bounded). For example, with respect to fieldbus communication, a formal analysis and suitable methodologies have been presented in [6], with the aim of guaranteeing before run-time that real-time distributed control systems can be successfully implemented with standard fieldbus communication networks.

There exist other approaches to monitor latencies and delays in distributed control systems based on wired component. The authors of [7] and [8] show two studies on modeling and analyzing latency and delay stability of network control systems. They evaluate fieldbus protocols and propose mechanism to mitigate latency and delays. The work [9] shows a model for end-to-end time delay dynamics in the Internet using system identification tools. The work in [10] presents an analytical performance evaluation of the switched Ethernet with multiple levels from timing diagram analysis and experimental evaluation from an experimental test bed with a networked control system.

4.2 Fieldbus Followers

Controller area network (CAN) [11] is an implementation of the fieldbus. It is a serial asynchronous multi-master communication protocol designed for applications needing high-level data integrity and data rates of up to 1 Mbit/s. CAN was originally developed by Bosch in the early 1980s for use in automobiles. It uses CSMA/CA for bus contention, which requires it to use an unbalanced, non-return-to-zero coding scheme, in this case RS232, for physical transmission. The publisher–subscriber model is used for information distribution. CAN is defined in ISO 11898 and only specifies the physical and data link layers. Due to its lack of high-level functionality, such as the provision of an application layer, CAN itself is unsuited for industrial automation. It is however used as the basis for other fieldbus protocols that define their own higher-level services above the CAN specification. Examples of such protocols include CANopen, DeviceNet, and ControlNet.

CANopen is a high-level expansion of CAN for use in automation. Several European industrial suppliers use it, and it was defined in the European EN 50325 standard, where a wide variety of application profiles are provided. They include motion control, building door control, and medical equipment.

ControlNet is an application layer extension of the CAN protocol and also defined in EN 50325. It implements the Common Industrial Protocol (CIP) application layer and is optimized for cyclical data exchange, making it more suited to process systems. As its name suggests, it was developed specifically for transmission of control data and has a high emphasis on determinism and strict scheduling. One notable feature of ControlNet is its built-in support for fully redundant cabling between devices. DeviceNet is a variant of ControlNet, with a focus on device-to-device communication.

The HART Communications Protocol (Highway Addressable Remote Transducer Protocol) [12] is another implementation of the fieldbus. It is a master/slave protocol, which means that a smart field (slave) device only speaks when spoken to by a master. The HART Protocol can be used in various modes, such as point-to-point or multi-drop, for communicating information to/from smart field instruments and central control or monitoring systems. In typical industrial systems, sensors communicate their signals using 4–20 mA analog cabled lines. The communication is done at 1,200 bps without interrupting the 4–20 mA signals, which allows a master to get two or more digital updates per second from a slave. Since it uses digital modulation (FSK), there is no interference with the 4–20 mA signals.

The PROFIBUS protocol [13] is another fieldbus implementation, arguably one of the most well-known and widely implemented fieldbuses, due to its endorsement by Siemens. It is composed of master and slave nodes. A master can send a message on its own initiative, once it receives the token, which circulates between masters in a logical ring fashion. Slaves do not have bus access initiative; they only acknowledge or respond to requests from masters. A transmission cycle comprises the request frame sent by an initiator (always a master) and the associated acknowledgment or response frame from the responder (usually a slave).

The PROFIBUS MAC protocol, being based on the measurement of the actual token rotation time, induces a well-defined timing behavior for the transferred messages, since the token cycle duration can be estimated prior to run-time [14, 15]. Both high- and low-priority messages are supported, as well as three acyclic data transfer services: Send Data with Acknowledgement (SDA), Send Data with No acknowledgement (SDN), and Send and Request Data (SRD).

Other variants are PROFIBUS Process Automation (PA), which is designed specifically for use in hazardous areas and is intrinsically safe; PROFIdrive for motion control; and PROFIsafe for safety systems [13]. All the variants implement a token-passing bus access strategy with multiple masters able to poll other devices for information, the main difference being the application profiles defined in each. This allows for a high degree of interoperability between the different buses. PROFIBUS is mainly implemented using RS485 at the physical layer, except for PROFIBUS PA, which makes use of the IEC 61158-2 physical layer to achieve intrinsic safety by limiting current on the bus.

PROFINET [16] is an application protocol that uses familiar concepts from PROFIBUS. However, it is implemented on the physical layer of the Ethernet. A proxy server is used to provide connectivity to PROFIBUS devices and additional protocols such as DeviceNet and Modbus. PROFINET real time is a progression of PROFINET which relies on the physical layer of the Ethernet but does not use either the TCP/IP or UDP/IP application layer. PROFIBUS uses information stored in the standard Ethernet frame header to identify packets that are real time. Network switches with Quality of Service (QoS) capabilities are then able to further prioritize packets. PROFINET makes use of remote procedure calls (RPC) and the distributed component object model (DCOM) for communications in the range of 50–100 ms, as well as modified Ethernet types for real-time communication.

Modbus [17] is an open master/slave application protocol that can be used on several different physical layers. It is an application layer messaging protocol, which provides client/server communication between devices connected on different types of buses or networks. Modbus-TCP means that the Modbus protocol is used on top of Ethernet-TCP/IP. Modbus-TCP is an open industrial Ethernet network protocol which has been specified by the Modbus-IDA User Organization in cooperation with the Internet Engineering Task Force (IETF) as an RFC Internet standard.

World Factory Instrumentation Protocol (WorldFIP) [18] was developed as an expansion of the original FIP protocol in an attempt to fulfill the requirements for an international fieldbus. WorldFIP supports two basic types of transmission services: exchange of identified variables and exchange of messages. The exchange of messages is used to support manufacturing message services, and the exchange of identified variables is based on a producer/consumer model, which relates producers and consumers within a distributed system. In order to manage any transactions associated with a single variable, a unique identifier is associated with each variable.

The WorldFIP data link layer (DLL) is made up of a set of produced and consumed buffers, which can be locally accessed (through application layer services) or remotely accessed through network services. In WorldFIP networks, the bus arbitrator table (BAT) regulates the scheduling of all buffer transfers. There are two types of buffer transfers that can be considered in WorldFIP. These are periodic and aperiodic (sporadic). The BAT imposes the schedule of the periodic buffer transfers and also regulates the aperiodic buffer transfers in the times that there are no periodic buffer transactions.

4.3 Wireless Industrial Networks

Wireless process control has been a popular topic in the field of industrial control [19–21]. Compared to traditional wired process control systems, wireless has a potential for cost savings, making installation easier (installation of wireless instruments in locations where cables may be restrictive, impractical, or vulnerable) and providing faster and simpler commissioning and reconfiguration (easy to replace and/or move in the plant site). Also, wireless technologies open up the potential for new automation applications. Wireless is also particularly suitable for installation on moving equipment where cabling may be easily damaged or restrict the operation of the machinery to be monitored. Faster commissioning and reconfiguration can also be done [20].

There have been some studies on hybrid fieldbus technology using IEEE 802.11 [22–25]. In [22, 23] the authors adapt the concept of fieldbus technology based on PROFIBUS to a hybrid setting (wired plus wireless). In [24, 25] the authors propose R-fieldbus, a wireless fieldbus protocol based on IEEE 802.11. According to [26], wireless fieldbus based on IEEE 802.11 has reliability limitations and incurs in high installation and maintenance costs.

The existing wireless technology was developed for use outside of industry and no considerations for real-time response or determinism are inherent in the medium. It also faces additional challenges that need to be addressed for industrial applications [20]: It is highly susceptible to interference from a variety of sources, which causes transmission errors. Within the transmission channel itself, effects such as multipath fading and inter-symbol interference are present. Environmental electromagnetic fields may also affect wireless transmission, such as those produced by large motors and electrical power transformers or discharges.

The limited distance over which wireless transceivers can operate, combined with the use of carrier sensing to determine when it is safe to transmit, may also result in problems, where two devices located out of the range of each other try and communicate with a third device that is located between them without knowledge of the other's actions. Wired carrier-sensing technologies such as the Ethernet are able to avoid such problems by ensuring that each device has knowledge of all others to which it is connected. Even with careful planning and device location, such knowledge cannot be guaranteed in a wireless medium. Wireless transceivers are also only able to operate at half duplex, as their own transmissions would overpower any signal they might be intended to receive.

Physical overhead on a wireless system is also significant in comparison to wired systems, as most wireless protocols require the transmission of predetermined data sequences before or during data transmission in order to evaluate and correct the effects of noise on the received information. Security of wireless transmission is also important, since physical access to the transmission medium cannot be restricted. Many wired fieldbuses are also able to make use of passively powered field devices, by supplying the energy required for the device's operation over the transmission medium. The existing wireless technologies have no such capability and provision for energy to remote devices is a concern, as is the energy efficiency of the remote devices.

In addition to difficulties in realizing general reliability and timeliness requirements, the characteristics of wireless transmission can negatively affect specific fieldbus methodologies. Fieldbuses often utilize unacknowledged transmission, since the probability of data not being received at all is relatively low. Such a strategy is unsuitable for wireless where the possibility of non-reception of a broadcast is significantly higher. This is especially troublesome in the case of token-passing networks, where the loss of the token may result in the bus needing to reinitialize to reestablish which device is the current master. Since interference is not uniform, some equipment may receive a broadcast while others do not. This can result in data inconsistency across a network in which the producer–consumer model is utilized.

Several techniques can be implemented to improve the performance of wireless in industrial application. Hidden node problems can be solved by adding a handshake system to the network, in which permission to transmit must be requested and granted before transmission may occur. However, this adds significant overhead to the channel, especially in the case of small data packets, where the initialization of transmission may require more time and data than the actual information to be communicated.

Each of the various technologies and mechanisms being investigated for wireless has its own advantages and disadvantages.

Several industrial organizations, such as HART [12], WINA [27], and ZigBee [28], have been pushing actively the application of wireless sensor technologies in industrial automation. Nowadays, it is possible to find WirelessHART [12] and ZigBee technologies and its protocols in those industrial applications. All of them are based on the IEEE 802.15.4 physical layer. IEEE 802.15.4 [29] is a standard which specifies the physical layer and media access control for low-rate wireless personal area networks (LR-WPANs).

WirelessHART [12] is an extension of wired HART, a transaction-oriented communication protocol for process control applications. To meet the requirements for control applications, WirelessHART uses TDMA technology to arbitrate and coordinate communications between network devices. The TDMA data link layer establishes links and specifies the time slot and channel to be used for communication between devices. WirelessHART has several mechanisms to promote network-wide time synchronization and maintains time slots of 10 ms length. To enhance reliability, TDMA is combined with channel hopping on a per-transaction (time slot) basis. In a dedicated time slot, only a single device can be scheduled for transmission in each channel (i.e., no spatial reuse is permitted).

ZigBee [28] is a specification for a suite of high-level communication protocols using small, low-power digital radios based on an IEEE 802.15.4 standard. ZigBee devices are often used in mesh network form to transmit data over longer distances, passing data through intermediate devices to reach distant ones. It is targeted at applications that require a low data rate, long battery lifetime, and secure networking. The basic channel access mode is CSMA/CA. However, ZigBee can send beacons on a fixed timing schedule which optimize the transmission and provide low latency for real-time requirements.

Those protocols are intended for use in communication with field instruments and fulfill a similar purpose to that of H1 fieldbuses. Although the terminology used to describe specific components differs from standard to standard, all of the standards are defined to cater for a similar set of devices. These are security and network management devices, gateway devices, routing devices, non-routing devices and handheld devices. The various instruments connect in a self-organizing hybrid star/mesh network, which is controlled by the network and security management devices. The management devices are powerful, wired devices, which interface to the wireless portion of the network through the gateway device. The gateway device can also be implemented as a protocol converter, making use of a wired fieldbus protocol to facilitate deterministic communication between the gateway and any controllers [30]. The mesh portion of the network is realized by the routing devices, which in turn connect nearby non-routing devices through the star portion of the network. Despite the similar operational philosophy of the protocols, they feature different network stacks and they are incompatible.

The implementation of wireless industrial networks will remain an active research area for a significant time, especially due to the fact that wireless communication is still developing and new technologies will need to be adapted for industrial applications.

4.4 Basic Considerations on Planning with Wireless Sensor Subnetworks

Generally, wireless sensor devices can be placed as they would have been placed in wired installations and connected to the rest of the wired installation by gateway devices. Typically, the gateway is placed first, since this is the core element of a network. There are three basic options for placing a gateway:

- Where it is easiest to integrate with the distributed control system or plant network.
- Central to the planned network. Placing the gateway in the center of the network provides the best position for most of the devices to have a direct communication link with the gateway.
- Central to the process unit. This placement provides the most flexibility for future expansion of the initially planned network to other areas within the process unit.

The wireless sensor devices themselves can be either directly connected to the gateway or form a wireless sensor network spanning a region and connecting to one of more gateways. According to [31], it is desirable to have at least 25 % of the wireless devices with a direct communication path to the gateway, since it ensures an ample number of data communication paths for the devices farther away. As for wired fieldbus planning, an operation schedule (cycle time) should be defined for the wireless sensor subnetworks within the industrial networks, in order to provide operations time planning. To develop the correct schedule, a static topology should be assumed, and the network must be sized according to timing requirements. This is necessary to avoid unwanted delays and latencies, which influences the overall system responsiveness.

While there are engineering guidelines and rules of thumb for planning operation and operation timings over wired industrial control systems, based on bus cycles and schedules, in a hybrid wired and wireless industrial control system with wireless sensor subnetworks, it is also necessary to define schedules for the wireless sensor subnetworks and to plan for operation timings end-to-end, over the whole system.

References

1. Flanagan B, Murray J (2003) Fieldbus in the cement industry. In: Proceedings of the cement industry technical conference 2003 conference record IEEEIASPCA 2003, pp 345–357
2. Goh HK, Devanathan R (2002) Fieldbus for control and diagnostics. In: Proceedings of the 7th international conference on control automation robotics and vision 2002 ICARCV 2002, vol 3, pp 1–21
3. Yokogawa Electric Corporation (2001) Technical information. In: Technical report Yokogawa Electric Corporation 2001, No. May
4. Fieldbus Foundation Members. Foundation fieldbus. [Online]. Available: http://www.fieldbus.org/
5. Wilson RL (1989) Smart transmitters-digital vs. analog. In: Proceedings of the forty-first annual conference of electrical engineering problems in the rubber and plastics industries
6. Gao H, Meng X, Chen T (2008) Stabilization of networked control systems with a new delay characterization. IEEE Trans Automat Control 53(9):2142–2148
7. Sato K, Nakada H, Sato Y (1988) Variable rate speech coding and network delay analysis for universal transport network. In: Proceedings of the IEEE INFOCOM 88 seventh annual joint conference of the IEEE computer and communications societies networks evolution or revolution
8. Wu J, Deng F-Q, Gao J-G (2005) Modeling and stability of long random delay networked control systems. In: Proceedings of the 2005 international conference on machine learning and cybernetics, vol 2, No. August, pp 947–952
9. Kamrani E, Mehraban MH (2006) Modeling internet delay dynamics using system identification. In: Proceedings of the 2006 I.E. international conference on industrial technology, No. c, pp 716–721
10. Lee KC, Lee S, Lee MH (2006) Worst case communication delay of real-time industrial switched ethernet with multiple levels. IEEE Trans Ind Electron 53(5):1669–1676
11. Bosch Controller Area (1998) Bosch Controller Area Network (CAN) version 2.0. In: Network, vol 1939
12. Hart Communication Foundation (2007) WirelessHART technical data sheet. In: ReVision, p 5
13. Xu J, Fang Y-J (2004) Profibus automation technology and its application in DP slave development. In: Proceedings of the international conference on information acquisition 2004
14. Zhang S, Burns A, Cheng T-H (2002) Cycle-time properties of the timed token medium access control protocol. IEEE Trans Comput 51(11):1362–1367
15. Li C, Tong W-M, Zhang Y-P (2009) Study on PROFIBUS frame transmission time property. In: Proceedings of the second international conference on information and computing science, vol 1
16. Ferrari P, Flammini A, Marioli D, Taroni A (2004) Experimental evaluation of PROFINET performance. In: Proceedings of the IEEE international workshop on factory communication systems, pp 331–334
17. Joelianto E, Hosana H (2008) Performance of an industrial data communication protocol on ethernet network. In: Proceedings of the 2008 5th IFIP international conference on wireless and optical communications networks WOCN 08, pp 1–5

18. Tovar E, Vasques F (2001) Distributed computing for the factory-floor: a real-time approach using WorldFIP networks. Comput Ind 44(1):11–31
19. Song J, Mok A, Chen D (2006) Improving PID control with unreliable communications. In: ISA EXPO technical report, No. October 2006, pp 17–19
20. Zheng L (2010) Industrial wireless sensor networks and standardizations: the trend of wireless sensor networks for process automation. In: Proceedings of the SICE annual conference 2010, pp 1187–1190
21. Lakkundi V, Beran J (2008) Wireless sensor network prototype in virtual automation networks. In: Proceedings of the First IEEE international workshop on generation C wireless networks
22. Willig A (2003) An architecture for wireless extension of PROFIBUS. In: Proceedings of the IECON03 29th annual conference of the IEEE industrial electronics society IEEE Cat No03CH37468, vol 3
23. Lee S, Lee KC, Lee MH, Harashima F (2002) Integration of mobile vehicles for automated material handling using Profibus and IEEE 802.11 networks. IEEE Trans Ind Electron 49(3):693–701
24. Rauchhaupt L (2002) System and device architecture of a radio based fieldbus-the RFieldbus system. In: Proceedings of the 4th IEEE international workshop on factory communication systems
25. Haehniche J, Rauchhaupt L (2000) Radio communication in automation systems: the R-fieldbus approach. In: Proceedings of the 2000 I.E. international workshop on factory communication systems proceedings Cat No00TH8531, pp 319–326
26. Choi D, Kim D (2008) Wireless fieldbus for networked control systems using LR-WPAN. J Control Automat Syst 6:119–125
27. WINA. Wireless industrial networking alliance. [Online]. Available: http://wina.org/. Accessed 22 Feb 2013
28. ZIGBEE. ZigBee Alliance > Home. [Online]. Available: http://www.zigbee.org/. Accessed 22 Feb 2013
29. Aliance I (2006) IEEE Std 802.15.4-2006. In: IEEE Std 8021542006 Revision of IEEE Std 8021542003, pp 0_1–305
30. Zhong T, Zhan M, Peng Z, Hong W (2010) Industrial wireless communication protocol WIA-PA and its interoperation with Foundation Fieldbus. In: Proceedings of the computer design and applications ICCDA 2010 international conference on, 2010, vol 4
31. Process Automation (2011) Technical white paper planning and deploying. In: Technical report of process automation

Chapter 5
Scheduling of Wireless Sensor Networks

This chapter discusses the state of the art related to scheduling and network planning within wireless sensor networks (WSN). It first provides some background information concerning medium access control (MAC), communication protocol approaches, and scheduling mechanisms used by time-division multiple access protocols (TDMA protocols). Planning mechanisms used to plan wireless sensor networks with precise timing are then examined.

5.1 Wireless Medium Access Control (MAC)

One key issue in WSNs that influences whether the deployed system will be able to provide timing guarantees is the MAC protocol and its configurations. The MAC protocols for wireless sensor networks can be classified broadly into two categories: contention- and schedule-based. The contention-based protocols can easily adjust to the topology changes as new nodes may join and others may die after deployment. These protocols are based on Carrier Sense Multiple Access (CSMA) mechanisms and have higher costs due to potential interference, message collisions, overhearing, and idle listening than the schedule-based counterparts. Schedule-based protocols can avoid those problems by defining precise schedules, but they have strict time synchronization requirements.

5.1.1 Contention-Based MAC Protocols

In contention-based MAC protocols, potential message receivers wake up periodically for a short time to sample the medium. When a sender has data, it transmits a series of short preamble packets, each containing the ID of the target node, until it either receives an acknowledgment (ACK) packet from the receiver or a maximum sleep time is exceeded. Following the transmission of each preamble packet, the transmitter node

J. Cecílio and P. Furtado, *Wireless Sensors in Industrial Time-Critical Environments*, Computer Communications and Networks, DOI 10.1007/978-3-319-02889-7_5,

waits for a timeout. If the receiver is not the target, it returns to sleep immediately. If the receiver is the target, it sends an ACK during the pause between the preamble packets. Upon reception of the ACK, the sender transmits the data packet to the destination.

These protocols are implemented using units of time called backoff periods. The expected number of times random backoff is repeated is a function of the probability of sensing the channel busy, which depends on the channel traffic. Since these do not provide a precise schedule to send data and use random backoff, these protocols have higher costs for message collisions, overhearing, and idle listening than schedule-based counterparts, and they are typically not used for applications requiring strict operation timing guarantees. On the other hand, they can easily adjust to topology changes, such as when new nodes join and others leave after deployment.

Some protocols frequently used in WSNs, such as S-MAC, B-MAC, WiseMAC, and X-MAC, are contention-based. S-MAC [1] defines periodic frame structure divided into two parts, with nodes being active in the first fraction of the frame and asleep for the remaining duration. The length of each of the frame parts is fixed, according to the desired duty cycle. Virtual clustering permits that nodes adopt and propagate time schedules, but it leads to the existence of multiple schedules, causing nodes at the border of more than one schedule to wake up more often. B-MAC [2] and WiseMAC [3] are based on Low-Power Listening (LPL) [2], that is, a very simple mechanism designed to minimize the energy spent in idle listening. Nodes periodically poll the medium for activity during a very short time, just enough to check if the medium is busy. If they find no activity, they return immediately to the sleep state for the rest of the period until the next poll. Nodes with data to send wake up the radio transmitting a long preamble (with minimum length equal to an entire poll period). This simple scheme can be quite energy-efficient in applications with sporadic traffic. However, the preamble size (which is inversely proportional to the desired duty cycle) must be carefully chosen not to be too large, since above a certain threshold it introduces extra energy consumption at the sender, receiver, and overhearing nodes, besides impairing throughput and increasing end-to-end latency. X-MAC [4] is also based on Low-Power Listening but reduces the overhead of receiving long preambles by using short and strobed preambles. This allows unintended receivers to sleep after receiving only one short preamble and the intended receiver to interrupt the long preamble by sending an ACK packet after receiving only one strobed preamble. However, even in X-MAC, the overhead of transmitting the preamble still increases with the wake-up interval, limiting the efficiency of the protocol at very low duty cycles.

5.1.2 *Schedule-Based MAC Protocols*

Schedule-based MAC protocols, such as time-division multiple access (TDMA), define a time schedule, which eliminates collisions and removes the need for a backoff. This increased predictability can better meet the requirements for timely data delivery. With a proper schedule, they allow nodes to get a deterministic access to the medium and provide delay-bounded services. TDMA is also power efficient, since it is inherently collision-free and avoids unnecessary idle listening, which are two major sources of energy consumption.

The main task in TDMA scheduling is to allocate time slots depending on the network topology and the node packet generation rates. TDMA is especially useful in critical real-time settings, where maximum delays must be provided. TDMA protocols will schedule the activity of the network in a period in which all nodes will be active. In the idle times between data gathering sessions, nodes can turn off the radio interface and lie in a sleep state.

The disadvantages of TDMA protocols are related with a lack of flexibility to modifications, such as adding more nodes, or data traffic changes over time. Another issue is that nodes have to wait for their own sending slot.

There are various works addressing TDMA protocols. Several protocols have been designed for quick broadcast/convergecast, others for generic communication patterns. The greatest challenges are the time slots, interference avoidance, low latencies, and energy efficiency. In RT-Link [5] protocol, time-slot assignment is accomplished in a centralized way at the gateway node, based on the global topology in the form of neighbor lists provided by the WSN nodes. It supports different kinds of slot assignment, depending on whether the objective function is to maximize throughput or to minimize end-to-end delay. Interference-free slot assignment is achieved by means of a 2-hop neighborhood heuristic, coupled with worst-case interference range assumptions.

WirelessHART [6] was designed to support industrial process and automation applications. In addition, WirelessHART uses at its core a synchronous MAC protocol called TSMP [7], which combines TDMA and Frequency Division Multiple Access (FDMA). The TSMP uses the benefits from synchronization of nodes in a multi-hop network, allowing scheduling of collision-free pair-wise and broadcast communication to meet the traffic needs of all nodes while cycling through all available channels.

GinMAC [8] is a TDMA protocol that incorporates topology control mechanisms to ensure timely data delivery and reliability control mechanisms to deal with inherently fluctuating wireless links. The authors show that under high traffic load, the protocol delivers 100 % of data in time using a maximum node duty cycle as little as 2.48 %. This proposed protocol is also an energy-efficient solution for time-critical data delivery with neglected losses.

PEDAMACS [9] is another TDMA scheme including topology control and routing mechanisms. The sink centrally calculates a transmission schedule for each node, taking interference patterns into account, and, thus, an upper bound for the message transfer delay can be determined. PEDAMACS is restricted by the requirement of a high-power sink to reach all nodes in the field in a single hop. PEDAMACS is analyzed using simulations, but a real-world implementation and corresponding measurements are not reported.

SS-TDMA [10] is a TDMA protocol designed for broadcast/convergecast in grid WSNs. The slot allocation process tries to achieve cascading slot assignments. Each node receives messages from the neighbors with their assigned slots. The receiving node knows the direction of an incoming message and adds a value to the neighbors slot number, in order to determine its own slot number. A distributed algorithm is required where each node is aware of its geometric position, limiting its applicability to grid topologies or systems where a localization service is available.

NAMA [11] is another TDMA protocol that tries to eliminate collisions dynamically: All nodes compute a common random function of the node identifier and of the time slot, and the node with the highest value is allowed to transmit in the slot. NAMA is based on a 2-hop neighborhood criterion (nodes at three hops of distance can reuse slots) for its decisions but presents an additional drawback of being computationally intensive.

5.2 Scheduling Mechanisms of TDMA Protocol

There exist a number of TDMA MAC protocols that are potentially suitable for use in wireless sensor networks. Since all of those protocols create a schedule for all network activity, in this subsection we describe, as an example, the scheduling mechanisms used by the GinMAC protocol [8]. The following description is based on [8, 12].

5.2.1 TDMA Dimensioning

A network dimensioning process is carried out before the network is deployed. The inputs for the dimensioning process are network and application characteristics that are known before deployment. The output of the dimensioning process is a TDMA schedule with epoch length E that each node has to follow.

The GinMAC TDMA epoch contains a number of basic slots, where each sensor can forward one message to the sink node and the sink node can transmit one broadcast message to each node for configuration or actuation. The GinMAC TDMA epoch also contains additional slots to improve transmission reliability. Finally, the GinMAC TDMA epoch may contain unused slots, which are purely used to improve the duty cycle of nodes. The three types of slots within the time frame must be balanced such that the delay, reliability, and energy consumption requirements are met.

The topology envelope in form of maximum hop distance H, fan-out degrees Oh $(0 < h \leq H)$, the maximum number of N_S^{Max} sensor nodes and N_A^{Max} actuator nodes is used to calculate the basic slot schedule within the GinMAC epoch. An example topology envelope for $H = 3$ and $O1 = 3$, $O2 = 3$, $O3 = 2$ is shown in Fig. 5.1. Basic slots are dimensioned under the assumption that all positions in the topology envelope will be occupied by nodes. For the N_A^{Max} actuators, a worst-case deployment is assumed, which means that they occupy places as far away from the sink as possible (leaf nodes).

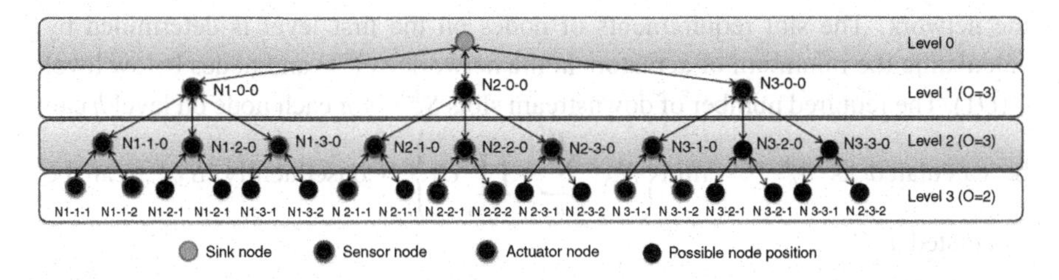

Fig. 5.1 Example of network topology with $H = 3, O1 = 3, O2 = 3, O3 = 2$, Na = 2 actuators and Ns = 15 sensors

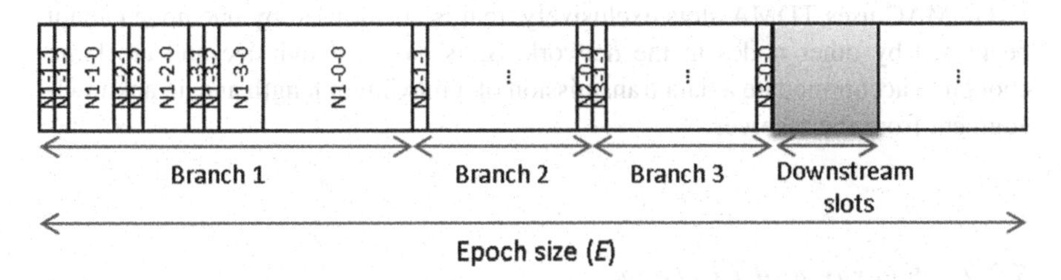

Fig. 5.2 GinMAC TDMA epoch

A number of slots S_{up} is required to accommodate traffic flowing from nodes (sensors and actuators) to the sink; a number of slots S_{down} is required to accommodate traffic flowing from the sink to actuators ($S = S_{\text{up}} + S_{\text{down}}$); S is the total number of basic slots required.

A leaf node (level OH) in the tree requires one basic TDMA slot within E to forward data to its parent node. This parent node requires a slot for each child node plus one slot for its own data for forwarding to its parent. Its slots must be located after the slots used by its children to ensure that data can travel within one GinMAC frame through the tree to the sink. The slot allocation for the previously given example is shown in Fig. 5.2. The GinMAC TDMA epoch includes slots for all nodes of Fig. 5.1. In the illustration of Fig. 5.2, we divided the network in three branches and represent only the full slot allocation for the first branch.

The total number of slots in E needed to forward data to the sink S_{up} can be calculated as follows. A node at tree level h requires $S_h^{\text{up}} = O_{h+1} \cdot S_{h+1}^{\text{up}} + 1$ with $S_h^{\text{up}} = 1$. Consequently, S_{up} can be calculated as

$$S_{\text{up}} = \sum_{h=1}^{H} S_h^{\text{up}} \cdot \prod_{i=1}^{h} O_i$$

The sink node must also be able to send commands to each actuator within one GinMAC epoch. Thus, the sink node requires as many slots as there are actuators in the network. The slot requirements of nodes on the first level is determined by calculating the minimum of actuators in the network (N_A^{Max}) and nodes below level 1 ($O1$). The required number of downstream slots S_h^{down} for each node on level h can be calculated as $S_h^{down} = \min\left\{N_A, \sum_{i=h+1}^{H}\prod_{j=h+1}^{i} O_i\right\}$. Consequently, S_{down} can be calculated as

$$S_{down} = S_0 + \sum_{h=1}^{H-1} S_h^{down} . \prod_{i=1}^{h} O_i$$

GinMAC uses TDMA slots exclusively, that is, a slot used by one node cannot be reused by other nodes in the network. Slots have a fixed size and are large enough to accommodate a data transmission of a maximum length and an acknowledgment from the receiver.

5.2.2 *Energy and Lifetime*

GinMAC can also include additional slots to improve the lifetime. The energy consumption of a node operating GinMAC can also be calculated before network deployment, based on the time the node remains awake and transmits, which is useful in deployments where maintenance schedules must be predictable.

For each node position in the topology, the worst-case and the best-case energy consumption can be determined. The worst case is incurred if nodes always transmit data in their available slot and if nodes make use of all available additional slots to improve transmission reliability. The best case is incurred if nodes do not transmit data, and only time synchronization messages have to be transmitted.

5.2.3 *Start of Node Operation*

After a node is switched on, it must first obtain time synchronization with the network. Data messages transmitted in the network are as well used to achieve time synchronization. The node continuously listens to overhear a packet from the sink node or a node that is already operating in the network. After overhearing one message, a node knows when the GinMAC frame starts, as each message carries information about the slot in which it was transmitted.

5.3 Brief Reference to Other Works on Wireless Sensor Network Scheduling

There are a number of scheduling algorithms for WSN networks [13–18]. In [15, 16] the authors review existing MAC protocols for WSNs that can be used in mission-critical applications. The reviewed protocols are classified according to data transport performance and suitability for those applications.

RAP [13] uses a velocity monotonic scheduling algorithm that takes into account both time and distance constraints. It maximizes the number of packets meeting their end-to-end deadlines, but reliability aspects are not addressed.

SPEED [14] maintains a desired delivery speed across the sensor network by a combination of feedback control and nondeterministic geographic forwarding. It is designed for soft real-time applications and is not concerned with reliability issues.

The Burst approach [17] presents a static scheduling algorithm that achieves both timely and reliable data delivery. This study assumes that a network topology is available and a deployment can be planned. Burst achieves end-to-end guarantees of data delivery in both the delay and reliability domains, and therefore it can support a mission-critical application.

The work in [18] describes the optimization problem of finding the most energy-preserving frame length in a TDMA system while still meeting worst-case delay constraints. The authors present an analytical approach to compute that value in generic sink trees. They also present an implementation using the existing DISCO Network Calculator framework [19].

Traffic regulation mechanisms are also explored as means to provide end-to-end guarantees using queuing models. In [20], the combination of queuing models and message scheduler turns into a traffic regulation mechanism that drops messages when they lose their expectations to meet predefined end-to-end deadlines.

The authors of [21] propose an energy-efficient protocol for low-data-rate WSNs. The authors use TDMA as the MAC layer protocol and schedule the sensor nodes with consecutive time slots at different radio states while reducing the number of state transitions. They also propose effective algorithms to construct data gathering trees to optimize energy consumption and network throughput.

In [22] the performance of SenTCP is examined, Directed Diffusion and HTAP, with respect to their ability to maintain low delays, to support the required data rates and to minimize packet losses under different topologies. The topologies used are simple diffusion, constant placement, random placement, and grid placement. It is shown that the congestion control performance, and consequently the packet delay, in sensor networks, can be improved significantly.

Latencies and delay modeling in wireless networks has been studied in previous works [23–25]. However, those works do not concern wireless sensor networks.

There are also some works addressing latency and delays for WSNs [26–29]. These works have considered the extension of the network calculus methodology [30] to WSNs. Network calculus is a theory for designing and analyzing deterministic queuing systems, which provides a mathematical framework based on min-plus and max-plus algebras for delay bound analysis in packet-switched networks.

In [26], the authors have defined a general analytical framework, which extends network calculus to be used in dimensioning WSNs, taking into account the relation between node power consumption, node buffer requirements, and the transfer delay. The main contribution is the provision of general expressions modeling the arrival curves of the input and output flows at a given parent sensor node in the network, as a function of the arrival curves of its children. These expressions are obtained by direct application of network calculus theorems. Then, the authors have defined an iterative procedure to compute the internal flow inputs and outputs in the WSN, node by node, starting from the lowest leaf nodes until arriving to the sink. Using network calculus theorems, the authors have extended the general expressions of delay bounds experienced by the aggregated flows at each hop and have deduced the end-to-end delay bound as the sum of all per-hop delays on the path. In [27], the same authors use their methodology for the worst-case dimensioning of WSNs under uncertain topologies. The same model of input and output flows defined by [26] has been used.

In [28], the authors have analyzed the performance of general-purpose sink-tree networks using network calculus and derived tighter end-to-end delay bounds.

In [29], the authors apply and extend the sensor network calculus methodology to the worst-case dimensioning of cluster-tree topologies, which are particularly appealing for WSNs with stringent timing requirements. They provide a fine model of the worst-case cluster-tree topology characterized by its depth, the maximum number of child routers, and the maximum number of child nodes for each parent router. Using network calculus, the authors propose "plug-and-play" expressions for the end-to-end delay bounds, buffering, and bandwidth requirements as a function of the WSN cluster-tree characteristics and traffic specifications.

End-to-end delay bounds for real-time flows in WSNs have been studied in [31]. The authors propose closed-form recurrent expressions for computing the worst-case end-to-end delays, buffering, and bandwidth requirements across any source-destination path in the cluster tree assuming error-free channel. They propose and describe a system model, an analytical methodology, and a software tool that permits the worst-case dimensioning and analysis of cluster-tree WSNs. With their model and tool, it is possible to dimension buffer sizes to avoid overflows and to minimize each cluster's duty cycle (maximizing nodes lifetime), while still satisfying messages deadlines.

In these works, the authors studied approaches to define a right scheduling to meet specific requirements, in particular, latency, within wireless sensor networks. They optimize the message path and data traffic to achieve their goals. In later chapters we describe solutions to plan operation timing for whole hybrid wired and wireless industrial control network. As in the previous works, slot-based planning is used for wireless sensor subnetworks, but the whole end-to-end network and operations are taken into account, defining operations schedules, predicting latencies, and dividing the network until latencies are according to the requirements.

References

1. Ye W, Heidemann J, Estrin D (2002) An energy-efficient MAC protocol for wireless sensor networks. In: Proceedings of the twenty-first annual joint conference of the IEEE computer and communications societies, IEEE, vol 3, No. c, pp 1567–1576
2. Polastre J, Hill J, Culler D (2004) Versatile low power media access for wireless sensor networks. In: SenSys'04 Proceedings of the 2nd international conference on embedded networked sensor systems, ACM, vol 1, pp 95–107
3. El-Hoiydi A, Decotignie JD (2004) WiseMAC: an ultra low power MAC protocol for the downlink of infrastructure wireless sensor networks. In: Proceedings ISCC 2004 ninth international symposium on computers and communications (IEEE Cat No04TH8769), IEEE, June–1 July 2004, vol 1, No. 28, pp 244–251
4. Buettner M, Yee GV, Anderson E, Han R (2006) X-MAC: a short preamble MAC protocol for duty-cycled wireless sensor networks. In: Proceedings of the 4th international conference on Embedded networked sensor systems, ACM, vol 76, No. May, pp 307–320
5. Rowe A, Mangharam R, Rajkumar R (2006) RT-Link: a time-synchronized link protocol for energy- constrained multi-hop wireless networks. In: Proceedings of the sensor and ad hoc communications and networks, 2006. SECON'06. 2006 3rd annual IEEE communications society on, 2006, IEEE, vol 2, No. C, pp 402–411
6. Hart Communication Foundation (2007) WirelessHART technical data sheet. Software Technologies Group, Inc [Online]. Available: http://www.stg.com/wireless/STG_Data_Sheet_WiHART_Software.pdf. Accessed 27 Nov 2013
7. Pister KSJ, Doherty L (2008) TSMP: time synchronized mesh protocol. In: Networks, vol 635, No. Dsn, pp 391–398
8. Suriyachai P, Brown J, Roedig U (2010) Time-critical data delivery in wireless sensor networks. Distrib Comput Sens Syst 6131:216–229
9. Ergen SC, Varaiya P (2006) PEDAMACS: power efficient and delay aware medium access protocol for sensor networks. IEEE Trans Mobile Comput 5(7):920–930
10. Kulkarni SS, Arumugam MU (2006) SS-TDMA: a self-stabilizing MAC for sensor networks. In: Phoha S, La Porta TF, Griffin C (eds) Sensor network operations. Wiley/IEEE Press, Hoboken/Piscataway, pp 1–32
11. Bao L, Garcia-Luna-Aceves JJ (2001) A new approach to channel access scheduling for ad hoc networks. In: MobiCom'01 Proceedings of the 7th annual international conference on mobile computing and networking, ACM, pp 210–221
12. Brown J, McCarthy B, Roedig U, Voigt T, Sreenan CJ (2011) BurstProbe: debugging time-critical data delivery in wireless sensor networks. In: Proceedings of the European conference on wireless sensor networks EWSN, ACM, vol 6, No. 2, pp 195–210
13. Lu C, Blum BM, Abdelzaher TF, Stankovic JA, He T (2002) RAP: a real-time communication architecture for large-scale wireless sensor networks. In: Proceedings eighth IEEE real time and embedded technology and applications symposium, vol 00, No. c, pp 55–66
14. Stankovic JA, Abdelzaher T (2003) SPEED: a stateless protocol for real-time communication in sensor networks. In: Proceedings of the 23rd international conference on distributed computing systems, IEEE, vol 212, No. 4494, pp 46–55
15. Suriyachai P, Roedig U, Scott A (2011) A survey of MAC protocols for mission-critical applications in wireless sensor networks. IEEE Commun Surv Tutor 14(2):240–264
16. Kredoii K, Mohapatra P (2007) Medium access control in wireless sensor networks. Comput Netw 51(4):961–994
17. Munir S, Lin S, Hoque E, Nirjon SMS, Stankovic JA, Whitehouse K (2010) Addressing burstiness for reliable communication and latency bound generation in wireless sensor networks. In: Proceedings of the 9th ACMIEEE international conference on information processing in sensor networks IPSN 10, ACM, No. May, p 303

18. Gollan N, Schmitt J (2007) Energy-efficient TDMA design under real-time constraints in wireless sensor networks. In: Proceedings of the 15th IEEE/ACM international symposium on modeling, analysis and simulation of computer and telecommunication systems (MASCOTS'07). IEEE
19. Gollan N, Zdarsky FA, Martinovic I, Schmitt JB (2008) The DISCO network calculator. In: Proceedings of the 14th GIITG conference on measurement modeling and evaluation of computer and communication systems MMB 2008
20. Karenos K, Kalogeraki V (2006) Real-time traffic management in sensor networks. In: Proceedings of the 2006 27th IEEE international real-time systems symposium RTSS06, 2006, vol 0, pp 422–434
21. Wu Y, Li X-Y, Liu Y, Lou W (2010) Energy-efficient wake-up scheduling for data collection and aggregation. IEEE Trans Parallel Distrib Syst 21(2):275–287
22. Vassiliou V, Sergiou C (2009) Performance study of node placement for congestion control in wireless sensor networks. In: Proceedings of the workshop of international conference on new technologies, mobility and security, IEEE, pp 3–10
23. Uhlemann E, Nolte T (2009) Scheduling relay nodes for reliable wireless real-time communications. In: Proceedings of the 2009 I.E. conference on emerging technologies factory automation, IEEE, pp 1–3
24. Yigitbasi N, Buzluca F (2008) A control plane for prioritized real-time communications in wireless token ring networks. In: Proceedings of the 2008 23rd international symposium on computer and information sciences, IEEE
25. Hou IH, Kumar P (2012) Real-time communication over unreliable wireless links: a theory and its applications. IEEE Wirel Commun 19(1):48–59
26. Schmitt J, Zdarsky F, Roedig U (2006) Sensor network calculus with multiple sinks. In: Proceedings of IFIP NETWORKING 2006 workshop on performance control in wireless sensor networks. ICST (Institute for Computer Sciences, Social-Informatics and Telecommunications Engineering), Brussels, Belgium. pp 6–13
27. Roedig U, Gollan N, Schmitt J (2007) Validating the sensor network calculus by simulations. Published in Proceeding WICON '07 Proceedings of the 3rd international conference on Wireless internet Article No. 34. ICST (Institute for Computer Sciences, Social-Informatics and Telecommunications Engineering), Brussels, Belgium. ISBN: 978-963-9799-12-7
28. Lenzini L, Martorini L, Mingozzi E, Stea G (2006) Tight end-to-end per-flow delay bounds in FIFO multiplexing sink-tree networks. Perform Eval 63(9–10):956–987
29. Koubaa A, Alves M, Tovar E (2006) Modeling and worst-case dimensioning of cluster-tree wireless sensor networks. In: Proceedings of the 2006 27th IEEE international real-time systems symposium RTSS06, IEEE, No. October, pp 412–421
30. D. Q. Systems (2004) NETWORK CALCULUS a theory of deterministic queuing systems for the internet. In: Online, vol 2050, pp xix–274. Online Version of the Book Springer Verlag – LNCS 2050 Version April 26, 2012
31. Jurcik P, Severino R, Koubaa A, Alves M, Tovar E (2008) Real-time communications over cluster-tree sensor networks with mobile sink behaviour. In: Proceedings of the 2008 14th IEEE international conference on embedded and real-time computing systems and applications, pp 401–412

Chapter 6
Latency Modeling for Distributed Control Systems with Wired and Wireless Sensors

In this chapter we will describe a latency model for end-to-end operation over hybrid wired and wireless industrial distributed control systems (DCS). Since nodes are configurable and the DCS may include actuators, the latency model can be decomposed in two parts: monitoring latency model (upstream) and commanding latency model (downstream).

We start by discussing monitoring latency. This latency corresponds to the latency that is measured from the sensing node to the control station. Then, Sect. 6.2 discusses the command latency model. This latency model is used to access the latency of sending a configuration command or the latency associated with an actuation command resulting from a closed-loop operation. Section 6.3 refers to closed-loop latencies and Sect. 6.4 discusses how to add non-real-time components to our monitor and command latency model. Finally, Sect. 6.5 discusses the prediction model used by the planning algorithm presented in the next chapter, used to dimension a DSC with timing guarantees.

6.1 Monitoring Latency Model

The monitoring latency can be divided into several parts (Fig. 6.1):

- The time elapsed between when an event happens and its detection by the node (t_{Event})
- The latency to acquire a sensor/event value (t_{Aq})
- The time needed to reach the transmission slot (this time can be neglected if we synchronize acquisition with the upstream slot for the sensor node) ($t_{WaitTXSlot}$)
- WSN latency ($t_{WSN_{UP}}$)
- Sink latency, which is the time needed to write messages coming from the WSN to the gateway (t_{Serial})
- Gateway latency, which corresponds to the time consumed to do some processing in the gateway ($t_{Gateway}$) (e.g., translation between protocols)

J. Cecílio and P. Furtado, *Wireless Sensors in Industrial Time-Critical Environments*, Computer Communications and Networks, DOI 10.1007/978-3-319-02889-7_6,

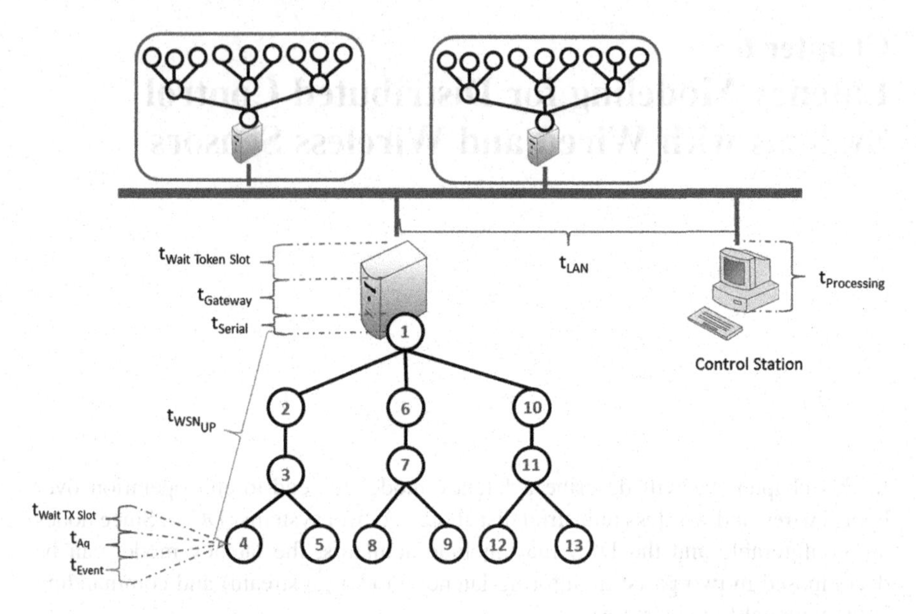

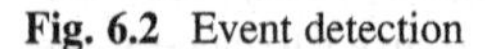

Fig. 6.1 Model for monitoring latency

Fig. 6.2 Event detection

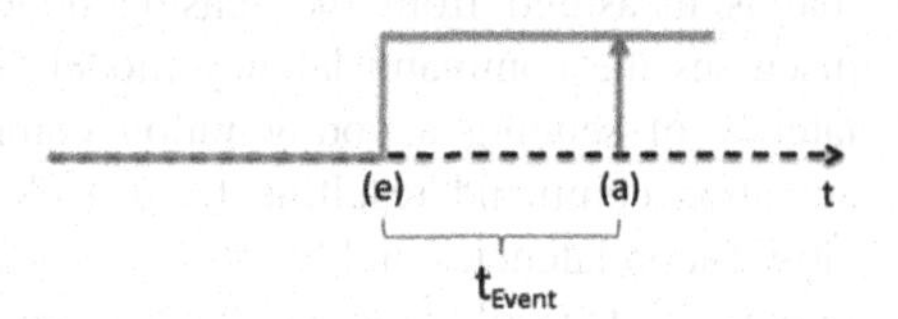

- The time needed to take the next token to deliver data to the control station ($t_{WaitTokenSlot}$)
- The local area network latency, which represents the latency needed to forward messages coming from the gateway to PLCs, computers, or control stations (t_{LAN})
- The latency associated with processing in the control station ($t_{\mathrm{Processing}}$)

Consider a reading going from a WSN node all the way to a control station. There is a latency associated with sensor sample (t_{Aq}), a latency associated with the elapsed time between sensor sample and a data transmission instant ($t_{WaitTXSlot}$), and a latency associated with the WSN path ($t_{WSN_{UP}}$), which corresponds to the time taken to transmit a message from a source node to the sink node.

If the instant when an external event manifests itself is considered, there is also a latency associated with the time elapsed between when the event first happened and its detection by the sensor node. As the event may occur in any instant (it is not synchronized with the sensor sampling and transmission slot), t_{Event} represents the amount of time between when an event manifests itself and when it is detected by the sampling mechanism. For instance, we are sampling temperature every second, and the event is a temperature above 90 °C. As shown in Fig. 6.2, if the event

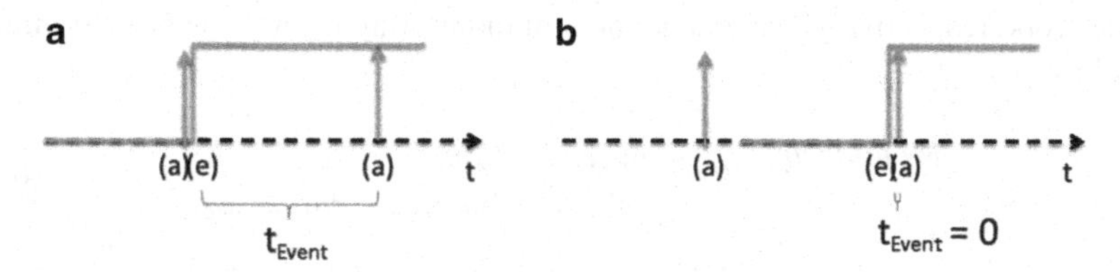

Fig. 6.3 Time diagram (from event to detection by the sampling mechanism) (**a**) maximum time, (**b**) minimum time

manifests itself for instant *e* onward and the next sampling instant is in instant *a*, then the wait time is t_{Event}.

Assuming a schedule-based protocol for a WSN subnetwork, t_{Event} can assume a value between zero and one epoch minus acquisition latency (t_{Aq}). As shown in Fig. 6.3a, if the event manifests itself immediately after the sampling instant, that event needs to wait one epoch minus acquisition latency to be detected. On the other hand (Fig. 6.3b), if the event manifests itself immediately before the sampling instant, t_{Event} is minimum and can be neglected.

Similar to t_{Event}, $t_{WaitTXSlot}$ varies from 0 to 1 epoch. It is minimum when sensor acquisition occurs immediately before transmission (acquisition and sending are synchronized) and maximum when sensor acquisition occurs after transmission slot (acquisition and sending are not synchronized).

In the gateway, there are there time intervals that can be considered (t_{Serial}, $t_{Gateway}$, and $t_{WaitTokenSlot}$). The first one (t_{Serial}) corresponds to the time needed by the sink node to write a message and gather it at the gateway side (e.g., if a serial interface is used, t_{Serial} is the time needed to write a message to the serial port, plus the time needed to read it by the gateway). The second component ($t_{Gateway}$) corresponds to the time needed for the gateway to get the message and do any processing that may have been specified over each message (e.g., translating the timestamps). The last part ($t_{WaitTokenSlot}$) corresponds to the time that a message must be kept in the gateway until a token reaches it, which allows transmitting that message.

The third part (t_{LAN}) corresponds to LAN transmission time, communication, and message handling software. Typically, this time is small because fast networks are used (fieldbus, Ethernet GigaBit) and have a significant bandwidth available. To simplify our model, we use $t_{Middleware}$ to refer to $t_{Gateway} + t_{LAN}$. Lastly, there is the time needed to perform the data analysis at control stations or other computer nodes ($t_{Processing}$).

Next we consider the latency for the three alternative network protocols configuration.

6.1.1 Wireless TDMA Plus Wired CSMA

Since the DCS may be built with different combinations of protocols, and assuming a TDMA and TCP/IP protocols for each wireless sensor subnetwork and wired subnetwork, respectively, the total amount of monitoring latency can be determined as follows:

$$Monitoring_{Latency} = t_{WSN_{AqE}} + t_{WSN_{UP}} + t_{Serial} + t_{Middleware} + t_{\mathrm{Processing}} \tag{6.1}$$

where $t_{WSN_{AqE}}$ represents the amount of latency introduced by the WSN subnetwork. It is given by

$$t_{WSN_{AqE}} = t_{Event} + t_{Aq} + t_{WaitTXSlot} \tag{6.2}$$

6.1.2 *Token-Based End-to-End*

On the other hand, when an end-to-end token-based approach is used, some extra latency is introduced. This extra latency is associated with the control station requesting the node to send data. In this case, the total monitoring latency can be determined as follows:

$$Monitoring_{Latency} = t_{Middleware} + t_{Serial} + t_{WSN_{Down}} + t_{CMD\mathrm{Processing}} + t_{WSN_{AqE}} + t_{WSN_{UP}} + t_{Serial} + t_{Middleware} + t_{\mathrm{Processing}} \tag{6.3}$$

where we have latency associated with requesting a data sent and latency associated with response.

6.1.3 *Wireless TDMA Plus Token-Based Wired*

Lastly, we may have hybrid solutions. For instance, if we have a TDMA protocol in wireless sensor subnetworks and a token-based protocol in wired part, when a message arrives at the gateway, it must wait ($t_{WaitforToken}$) until the token reaches that gateway, which allows it to transmit its data to the control station. In this case, the total amount of monitoring latency can be determined as follows:

$$Monitoring_{Latency} = t_{WSN_{AqE}} + t_{WSN_{UP}} + t_{WaitforToken} + t_{Middleware} + t_{CMD\mathrm{Processing}} + t_{Serial} + t_{Middleware} + t_{\mathrm{Processing}} \tag{6.4}$$

6.2 Command Latency Model

In the down path, used by configuration or actuation commands, there are also latency parts that can be identified. Figure 6.4 shows those parts. The model is defined for TDMA, token-based, and hybrid strategies.

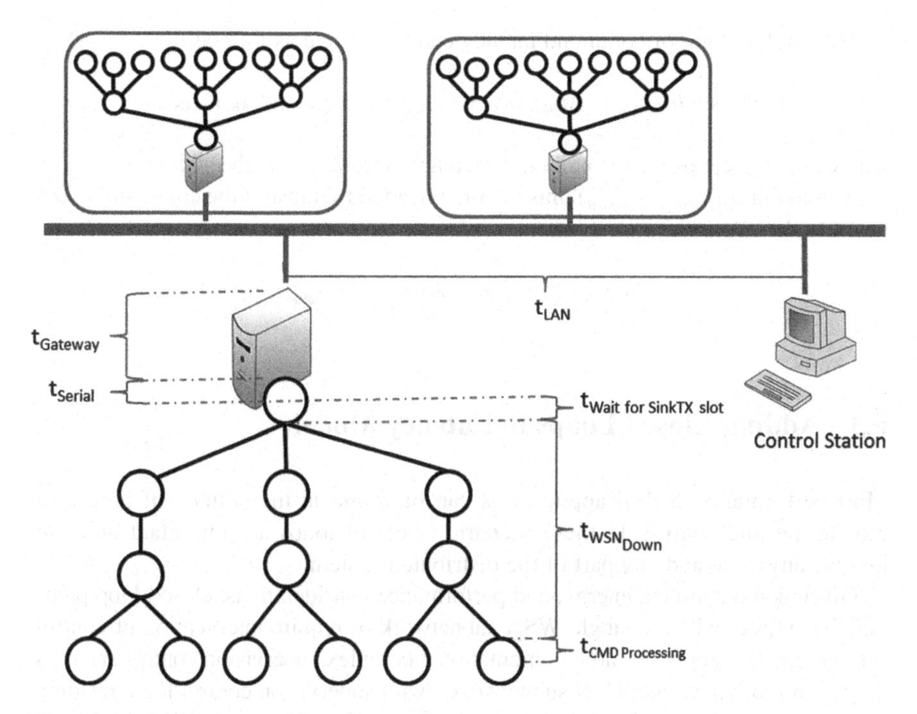

Fig. 6.4 Model for command latency

Consider a command sent from a control station to a node. Similar to upstream data transmission, there is LAN transmission latency (t_{LAN}). In the gateway, there are three time intervals that can be considered ($t_{Gateway}$, t_{Serial}, and $t_{WaitforSinkTXslot}$). $t_{Gateway}$ and t_{Serial} are similar to upstream data latency. $t_{Gateway}$ corresponds to the time needed for the gateway to receive the command, do any processing, and send it to the serial port. t_{Serial} corresponds to the time needed by the sink node to read the command from the serial interface. If the WSN subnetwork is running a TDMA schedule, upon receiving the command by the sink node, it needs to wait $t_{WaitforSinkTXslot}$ to reach the transmission slot to send the command to the target node. This latency part represents the amount of time that a command is kept in the sink node until it gets a downstream slot.

Since the WSN subnetwork schedule is planned, it is possible to place downstream slots in specific positions in the schedule to have a reasonable small value for this latency, but since the instant of a command is submitted may not be predictable, $t_{WaitforSinkTXslot}$ may be large.

If the WSN subnetwork is running a CSMA protocol, upon receiving the command by the sink node, it is sent to the target node immediately. In this case, the $t_{WaitforSinkTXslot}$ will be zero.

Lastly, there are latencies associated with the WSN path ($t_{WSN_{Down}}$) and the target node processing the command ($t_{CMD\text{Processing}}$). $t_{WSN_{Down}}$ corresponds to the time taken to transmit a command from the sink node to the target node, while $t_{CMD\text{Processing}}$ corresponds to the time taken to process the command inside the target node.

The total amount of command latency can be defined as follows:

$$Command_{Latency} = t_{Middleware} + t_{Serial} + t_{WSN_{CMD}} + t_{CMD\text{Processing}} \tag{6.5}$$

where $t_{WSN_{CMD}}$ represents the amount of latency introduced by the sink node to send the command ($t_{WaitforSinkTXslot}$) plus the time needed to transmit the command to the target node ($t_{WSN_{Down}}$).

$$t_{WSN_{CMD}} = t_{WaitforSinkTXslot} + t_{WSN_{Down}} \tag{6.6}$$

6.3 Adding Closed Loops to Latency Model

Most performance-critical applications can be found in the domain of industrial monitoring and control. In these scenarios, control loops are important and can involve any node and any part of the distributed system.

Given computational, energy, and performance considerations, closed-loop paths may be entirely within a single WSN subnetwork or require intervention of control station (e.g., for applying more computational complex supervision controller), or it may span more than one WSN subnetworks, with supervision control logic residing in one of the distributed PLC outside the WSNs (middleware servers).

The closed-loop latency is the time taken from sensing node to the actuator node, passing through supervision control logic. It will be the time taken since the sample is gathered in the sensing node to the instant when the action is performed in the actuator node.

The position of supervision control logic may depend on timing restrictions and data needed to compute decisions. Given latency considerations, closed-loop paths may be entirely within a single WSN subnetwork or require intervention of control station (e.g., for applying more computational complex supervision controller), or it may span more than one WSN subnetworks, with supervision control logic residing in one of the distributed nodes outside the WSNs (middleware servers). Moreover, the closed-loop decision may be asynchronous (when a data message from sensing nodes participating in the decision arrives) or synchronous (at defined time periods).

Asynchronous control can be defined as "upon reception of a data message, the supervision control logic computes a command and sends it to an actuator." The control computations are based on events. For instance, we can configure a node to send data messages only if a certain threshold condition was met.

Synchronous control can be defined as a periodic control, where the periodicity of computations is defined by users. The supervision control logic runs in specific instants (period). Typically, this type of controller involves multiple samples for the computation of the actuation commands, and multiple nodes may participate in sensing or actuation.

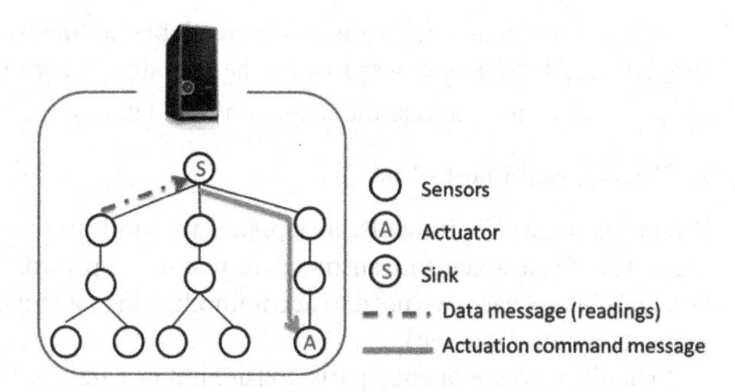

Fig. 6.5 Control decision at sink node

6.3.1 Supervision Control Logic in the Gateway

The closed-loop latency can be estimated for each protocol alternatives (wireless TDMA plus wired CSMA, token-based end-to-end and wireless TDMA plus Token-based wired) and for each strategy (asynchronous, synchronous).

Figure 6.5 shows a scenario example of closed-loop system where the supervision control logic is implemented in the sink node.

At the sink node, when a data message from sensing nodes participating in the decision arrives (asynchronous control) or at defined time periods (synchronous control), the condition and thresholds are analyzed, and the actuator is triggered if one of the defined conditions is matched.

1. Wireless TDMA plus wired CSMA

In this alternative, data readings must be acquired and sent from sensor node to the sink node ($t_{WSN_{AqE}}$), transferred from sink node to the gateway (t_{Serial}), and processed in the gateway to determine the actuation command ($t_{\text{Processing}}$). After concluding the processing, the resulting command is sent to an actuator. So, the command must be written in the serial interface to be received by the sink node (t_{Serial}), which will keep it until it gets the downstream slot ($t_{WaitTXSlot}$), and, when it arrives, the sink will forward the command to the actuator ($t_{WSN_{CMD}}$). Lastly, the command is processed by the target node ($t_{CMD\text{Processing}}$).

The closed-loop latency for asynchronous control for wireless TDMA plus wired CSMA network can be estimated by

$$\begin{aligned} CL_{Latency_{Async}} = & \; t_{WSN_{AqE}} \\ & + t_{Serial} + t_{\text{Processing}} + t_{Serial} + t_{WaitTXSlot} \\ & + t_{WSN_{CMD}} + t_{CMD\text{Processing}} \end{aligned} \tag{6.7}$$

Concerning synchronous control, the closed-loop latency for gateway decisions can be estimated as

$$\begin{aligned} CL_{Latency_{Sync}} = & \; t_{\text{Processing}} + t_{Serial} + t_{WaitTXSlot} \\ & + t_{WSN_{CMD}} + t_{CMD\text{Processing}} \end{aligned} \tag{6.8}$$

We assume that data values are available at the supervision control logic (implemented in the gateway) when the decision is computed. In this case, the $CL_{Latency_{Sync}}$ does not include the upstream data latency.

2. Token-based End-to-End

When token-based protocols are applied to wireless sensor networks, each node must wait for a token to transmit data readings upward. To estimate closed-loop latencies in this case, we need to accommodate the latency associated with the node request (sending a token).

In addition to the latency parts considered in wireless TDMA plus wired CSMA alternative, we must add an extra time to write the token to the sink node (t_{Serial}), to send it to the target node ($t_{WSN_{CMD}}$), and to process it at the target node ($t_{CMD\text{Processing}}$). In this alternative, the closed-loop latency for asynchronous control can be estimated by

$$\begin{aligned} CL_{Latency_{Async}} &= t_{Serial} + t_{WaitTXSlot} + t_{WSN_{CMD}} + t_{CMD\text{Processing}} \\ &+ t_{WSN_{AqE}} + t_{Serial} + t_{\text{Processing}} \\ &+ t_{Serial} + t_{WaitTXSlot} + t_{WSN_{CMD}} + t_{CMD\text{Processing}} \end{aligned} \tag{6.9}$$

In Eq. 6.8, $t_{WaitTXSlot}$ can be neglected because we assume a CSMA network where each node "talks" only when it has the token. There is a single token in the network. In the synchronous control, the closed-loop latency is given by Eq. 6.8, but, similar to asynchronous estimation, $t_{WaitTXSlot}$ can be neglected.

3. Wireless TDMA plus Token-based wired

Since we are assuming a wireless TDMA and the closed-loop decision is done at the gateway, this third alternative is equal to the first alternative. So, closed-loop latency for asynchronous control for wireless TDMA plus token-based wired network can be estimated by Eq. 6.7 and, in the synchronous case, can be estimated by Eq. 6.8.

6.3.2 *Supervision Control Logic in the Control Station*

The other alternative for closed-loop control is to deploy the supervision control logic in one control station. In this case, the control loop operation may involve parts of the network with one network protocol (e.g., TDMA) and others with fieldbus protocols.

This alternative is also shown in Fig. 6.6. In this case it is possible to read data from one or more WSNs, to compute a decision based on more complex algorithms, and to actuate over the distributed system. The closed-loop algorithm will receive data coming from sensors and will produce actuation commands for the actuator(s).

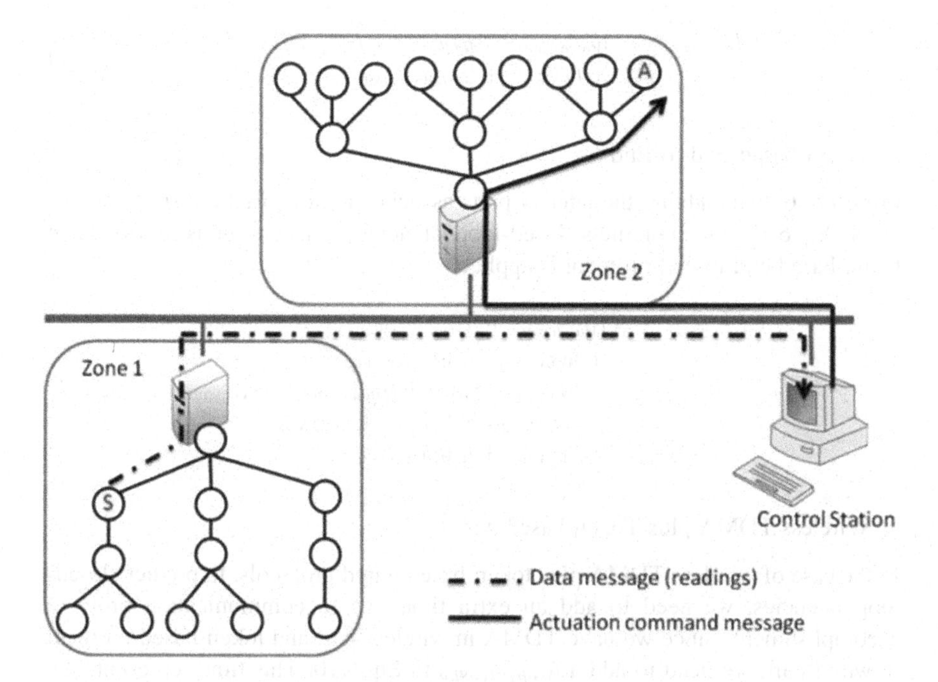

Fig. 6.6 Closed loop over whole distributed system

When the supervision control logic is deployed in a control station, the latency model must accommodate latency parts associated with the time needed to transfer data and commands in both parts of the network (wireless and wired parts). So, to estimate the closed-loop latency, and after defining the latency model to the gateway option, we need to add the corresponding part of latency due to the wired part. Next, we extend the latency model defined to the supervision control logic implemented in a gateway.

1. Wireless TDMA plus wired CSMA

From Eq. 6.7 and adding $t_{Middleware}$, which corresponds to the time needed to transmit a message from the gateway to the control station and do the message handling, the closed-loop latency for asynchronous control can be estimated by

$$\begin{aligned} CL_{Latency_{Async}} = & \; t_{WSN_{AqE}} \\ & + t_{Serial} + t_{Middleware} + t_{\text{Processing}} \\ & + t_{Middleware} + t_{Serial} + t_{WaitTXSlot} \\ & + t_{WSN_{CMD}} + t_{CMD\text{Processing}} \end{aligned} \tag{6.10}$$

Extending Eq. 6.8 to the control station option, we obtain Eq. 6.11 which predicts closed-loop latencies for synchronous control.

$$CL_{Latency_{Sync}} = t_{\text{Processing}} + t_{Middleware} + t_{Serial} + t_{WaitTXSlot} + t_{WSN_{CMD}} + t_{CMD\text{Processing}} \tag{6.11}$$

2. Token-based End-to-End

From Eq. 6.9 and adding the latency parts associated with wired transmission, we obtain Eq. 6.12, which predicts closed-loop latencies for synchronous control when token-based end-to-end protocol is applied:

$$\begin{aligned} CL_{Latency_{Async}} = {} & t_{Middleware} + t_{Serial} + t_{WaitTXSlot} \\ & + t_{WSN_{CMD}} + t_{CMD\text{Processing}} \\ & + t_{WSN_{AqE}} + t_{Serial} + t_{Middleware} + t_{\text{Processing}} \\ & + t_{Middleware} + t_{Serial} + t_{WaitTXSlot} \\ & + t_{WSN_{CMD}} + t_{CMD\text{Processing}} \end{aligned} \tag{6.12}$$

3. Wireless TDMA plus Token-based wired

In the case of wireless TDMA plus token-based wired protocols, to predict closed-loop latencies, we need to add an extra time due to communication protocol accomplishment. Since we have TDMA in wireless part and token-based protocol in wired part, we need to add a $t_{WaitTokenSlot}$ to Eq. 6.10. This time represents the time needed to take the next token to deliver data to the control station:

$$\begin{aligned} CL_{Latency_{Async}} = {} & t_{WSN_{AqE}} \\ & + t_{Serial} + t_{WaitTokenSlot} + t_{Middleware} + t_{\text{Processing}} \\ & + t_{Middleware} + t_{Serial} + t_{WaitTXSlot} \\ & + t_{WSN_{CMD}} + t_{CMD\text{Processing}} \end{aligned} \tag{6.13}$$

6.4 Adding Non-real-Time Components

We assume non-real-time cabled components, such as gateways, PLCs, and computers running operating systems such as Windows, Unix, or Linux and communicating using TCP/IP. Those components are responsible for part of the end-to-end latencies. Those latencies must be characterized by system testing. This is done by running the system with an intended load, while collecting and computing latency statistics.

Next we show two setups (Figs. 6.7 and 6.8 corresponding to a small and a larger network, respectively) that can be used to characterize latencies in those parts. The first example consists of a small network (Fig. 6.7). It includes 13 TelosB mote and 2 computers connected through a wired Ethernet. A TelosB is attached to a computer (gateway) via serial interface. It receives data messages from other nodes and writes them in the serial interface. Each node generates a data message per second. The gateway computer has a dispatcher which forwards each message to

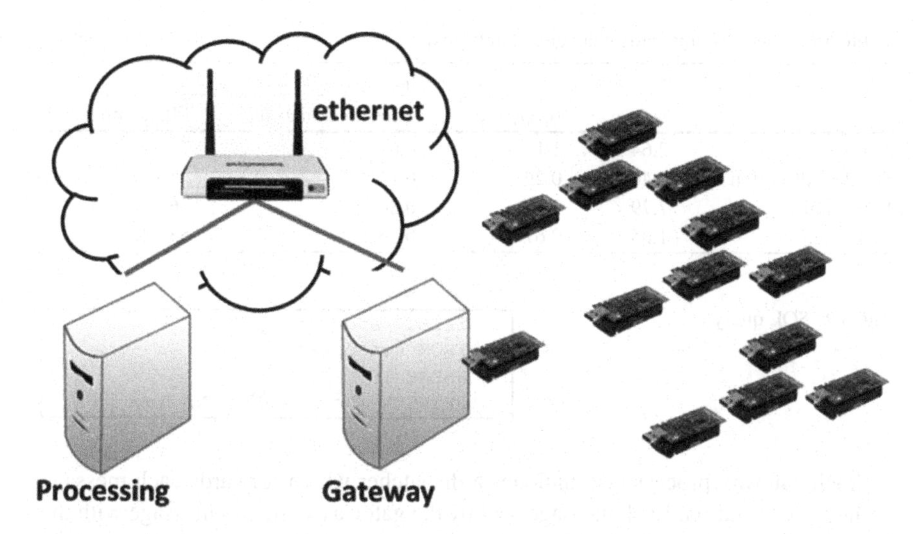

Fig. 6.7 Testing setup

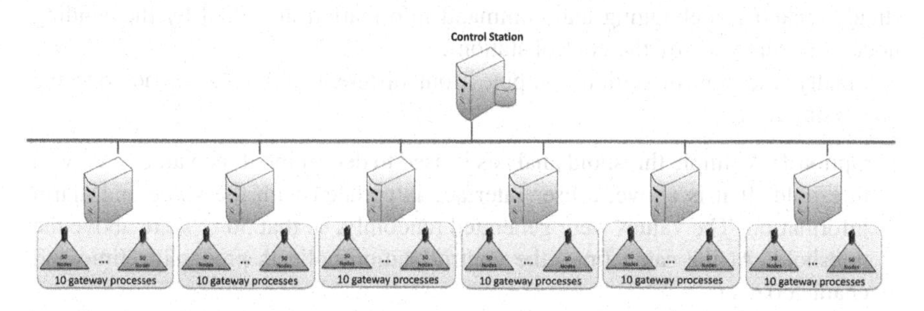

Fig. 6.8 Distributed control system – testing setup

the processing computer. Finally, the processing computer computes two different alternative operations to characterize the processing time: a simple threshold and a more complex algorithm to compute the PID gains and the actuation commands.

Table 6.1 shows the latency characterization for this setup. All times are given in milliseconds.

The most important measure in Table 6.1 is the maximum value. This value allows bounding the latency. In this example, latency would be bounded by (7.79 + 3.14 + 0.86) milliseconds in the case of threshold analysis operation and (7.79 + 3.14 + 86.22) milliseconds for the PID controller.

The second example (Fig. 6.8) consists of a distributed control system with 3,000 sensors. The setup includes 3,000 sensors, 6 gateway computers, and a control station. All computers are connected through a wired network. Each WSN subnetwork is composed of 50 nodes, and each gateway computer has 10 gateway processes running. Each node generates a data message per second.

Table 6.1 Non-real-time parts characterization [ms]

	t_{Serial}	$t_{Middleware}$	$t_{Processing}$ Threshold analysis	$t_{Processing}$ PID computation
Average	2.64	1.12	0.61	73.61
Standard deviation	0.40	0.29	0.14	5.14
Maximum	7.79	3.14	0.86	86.22
Minimum	1.85	0.67	0.36	53.62

Fig. 6.9 SQL query

```
SELECT avg(temp), TS
FROM sensorData
WHERE TS
BETWEEN (now-60000) AND now
```

Each gateway process also includes a dispatcher which forwards each message to the control station. Each message sent by the gateway is an xml message with the format: sensor values (temperature, humidity, light), timestamps (generation time, dispatcher in, dispatcher out), performance, debugging, and command information (fields related to debugging and command information are filled by the sending node to be analyzed by the control station).

Finally, the control station computes four different things to characterize the processing time:

- Option 1: A simple threshold analysis is used to determine if the value is above a threshold. If it is above, a user interface is updated with the value and alarm information. The values were generated randomly, so that 50 % were above the threshold. At the same time, the control station collects processing time and characterizes it.
- Option 2: An average of last 50 samples per node is computed to compare the result with the threshold value. Similar to the previous case, if value was greater than the threshold, an alarm is generated.
- Option 3: Insert into a database. Each message that arrives at the control station is stored in a database without further processing.
- Option 4: Insert into a database and request the database to compute the average of last received messages per node. Each message that arrives at the control station is stored in a database. After that, the control station submits an SQL query (Fig. 6.9) to the database to compute the average temperature of messages received in the last 60 s.

Table 6.2 shows the non-real-time parts characterization for this setup. All times are given in milliseconds.

In this example, and assuming the same serial latency as in Table 6.1, the latency of message stored in the database would be bounded by (7.79 + 5.25 + 8.31) milliseconds.

Table 6.2 Non-real-time parts characterization, second setup [ms]

	$t_{Middleware}$	$t_{\text{Processing}}$			
		Option 1	Option 2	Option 3	Option 4
Average	3.50	2.56	3.43	7.67	12.39
Standard deviation	0.87	0.51	0.48	1.52	1.21
Maximum	5.25	5.33	5.75	8.31	18.06
Minimum	1.75	0.23	0.27	7.04	10.80

6.5 Prediction Model for Maximum Latency

In the previous section, we discuss the base latency model for three different configurations of the network (wireless TDMA plus wired CSMA, token-based end-to-end, and wireless TDMA plus token-based wired). The presented model is useful to understand which parts of latencies are involved in each operation (monitoring and actuation). However, when we want to predict the latency for an operation, we need to determine the maximum bound of that latency. In this section, we will discuss the maximum bound for each latency part, which will allow predicting operation latencies.

6.5.1 Wireless TDMA Plus Wired CSMA

Equations 6.1 and 6.5 represent monitoring and commanding latencies. However, maximum values for each latency should be used because we are considering strict end-to-end operation timing guarantees. In this subsection we discuss how to predict maximum operation latencies and how to obtain the maximum value for each part of the latency.

The maximum monitoring latency for the combination of TDMA–CSMA can be determined as follows:

$$\max\left(Monitoring_{Latency}\right) = \max\left(t_{WSN_{AqE}}\right) + \max(t_{WSN_{UP}}) + \max(t_{Serial}) + \max(t_{Middleware}) + \max\left(t_{\text{Processing}}\right) \quad (6.14)$$

The maximum (bounds) for some parts, such as t_{Serial}, $t_{Middleware}$, and $t_{\text{Processing}}$, may be assumed or characterized experimentally. The maximum value of $t_{WSN_{AqE}}$ can be predicted by analyzing the maximum values of its parts (Eq. 6.15).

$$\max\left(t_{WSN_{AqE}}\right) = \max(t_{Event}) + \max\left(t_{Aq}\right) + \max(t_{WaitTXSlot}) \quad (6.15)$$

The values of t_{Event} and $t_{WaitTXSlot}$ were described in the previous section. Assuming that the network is configured to sense and send a data item per epoch only, t_{Event} and $t_{WaitTXSlot}$ can vary from 0 to 1 epoch size, where the maximum value for each is an epoch size.

Therefore, $t_{WSN_{AqE}}$ can assume one of the following four alternatives:

- $\max(t_{WSN_{AqE}}) = \max(t_{Aq}) + \max(t_{WSN_{UP}})$ – when the acquisition instant is synchronized with the sending instant and we do not consider the event start instant
- $\max(t_{WSN_{AqE}}) = \max(t_{Aq}) + \max(t_{WaitTXSlot}) + \max(t_{WSN_{UP}})$ – when the acquisition instant is not synchronized with the sending instant and we do not consider the event start instant
- $\max(t_{WSN_{AqE}}) = \max(t_{Event}) + \max(t_{Aq}) + \max(t_{WSN_{UP}})$ – when the acquisition instant is synchronized with the sending instant and we are considering the event start instant
- $\max(t_{WSN_{AqE}}) = \max(t_{Event}) + \max(t_{Aq}) + \max(t_{WaitTXSlot}) + \max(t_{WSN_{UP}})$ – when the acquisition instant is not synchronized with the sending instant and we are considering the event start instant

Similar to non-real-time parts, t_{Aq} must be characterized by experimental evaluation. It depends on the node platform and which sensor is sampled. The maximum value results from the collection of time measures during the experiment.

Lastly, $\max(t_{WSN_{UP}})$ depends on the node position in the WSN topology and how data flows in the network.

We assume that data can be forwarded using one of the following alternatives:

- Each node sends data to its parent. Each parent receives data from a child and forwards it immediately to its parents.
- Each node sends data to its parent. Each parent collects data from all children and only forwards up after receiving from all children.
- Each node sends data to its parent. Each parent collects data from all children, aggregates data messages from all children, and only forwards a merged message to its parents.

Depending on which alternative is used, $\max(t_{WSN_{UP}})$ can be predicted. For instance, assuming the WSN subnetwork shown in Fig. 6.10, the first alternative to define the data flow is used, and in maximum, 10 ms is needed to transmit a message between two consecutive nodes, $\max(t_{WSN_{UP}})$ for a node at level 3 (e.g., node 4 or 5) is 30 ms.

6.5.2 Token-Based End-to-End

In Eq. 6.3, we show which parts of latency are considered to determine the end-to-end latency of a monitoring operation when a token-based protocol is used. To predict

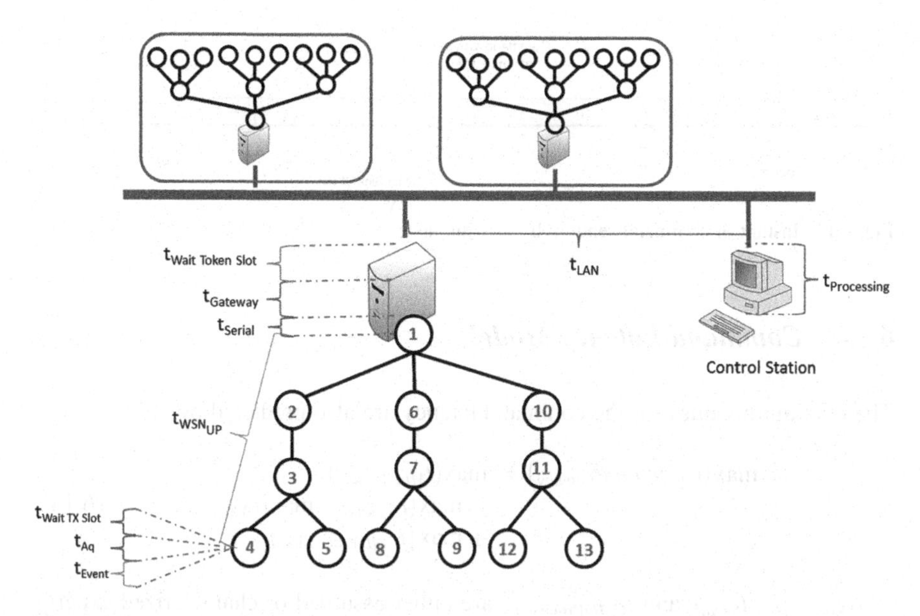

Fig. 6.10 Model for monitoring latency

that latency, the maximum value of each part must be considered. So, the maximum amount of monitoring latency can be determined as follows:

$$\begin{aligned}\max\left(Monitoring_{Latency}\right) = {} & \max(t_{Middleware}) + \max(t_{Serial}) + \max(t_{WSN_{Down}}) \\ & + \max\left(t_{CMD\mathrm{Processing}}\right) + \max\left(t_{WSN_{AqE}}\right) + \max(t_{WSN_{UP}}) \\ & + \max(t_{Serial}) + \max(t_{Middleware}) + \max\left(t_{\mathrm{Processing}}\right)\end{aligned} \quad (6.16)$$

6.5.3 *Wireless TDMA Plus Token-Based Wired*

Similar to the previous definitions, the maximum amount of monitoring latency when a TDMA protocol is used in wireless sensor subnetwork and when a token-based protocol is used in wired parts can be determined based on the maximum values of each part represented in Eq. 6.4. It is determined as follows:

$$\begin{aligned}\max\left(Monitoring_{Latency}\right) = {} & \max\left(t_{WSN_{AqE}}\right) + \max(t_{WSN_{UP}}) + \max\left(t_{WaitforToken}\right) \\ & + \max(t_{Middleware}) + \max\left(t_{CMD\mathrm{Processing}}\right) \\ & + \max(t_{Serial}) + \max(t_{Middleware}) + \max\left(t_{\mathrm{Processing}}\right)\end{aligned} \quad (6.17)$$

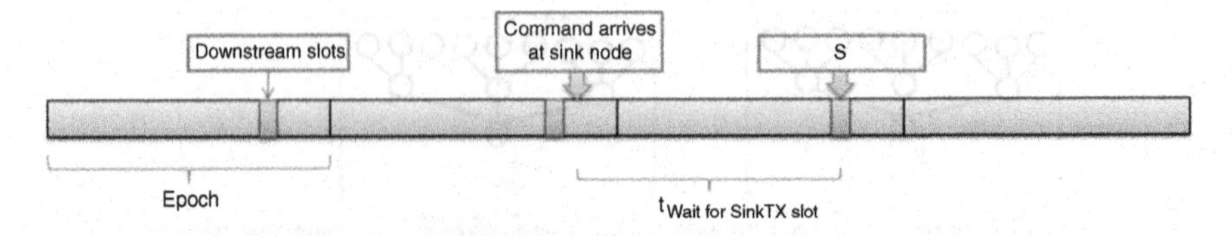

Fig. 6.11 Instant of command sending by the sink node

6.5.4 Command Latency Model

The maximum values for the command latency are also predicted by

$$\begin{aligned}\max\left(Command_{Latency}\right) = {} & \max(t_{Middleware}) \\ & + \max(t_{Serial}) + \max(t_{WSN_{CMD}}) \\ & + \max\left(t_{CMD\text{Processing}}\right)\end{aligned} \tag{6.18}$$

$t_{Middleware}$, t_{Serial}, and $t_{CMD\text{Processing}}$ are either assumed or characterized experimentally, while $\max(t_{WSN_{CMD}})$ is determined by Eq. 6.19.

$$\max(t_{WSN_{CMD}}) = \max\left(t_{WaitforSinkTXslot}\right) + \max(t_{WSN_{Down}}) \tag{6.19}$$

The value $\max(t_{WaitforSinkTXslot})$ is an epoch size. For instance, we are sending an actuation command to a WSN node. As shown in Fig. 6.11, if the command arrives immediately after the sending instant and the next sending instant is in instant s, then the wait time is $t_{WaitforSinkTXslot}$. If only one downstream slot was provided per epoch, $\max(t_{WaitforSinkTXslot}) = epoch\ size$.

This can be modified by adding equally spaced downstream slots in the schedule, resulting in a $\max\left(t_{WaitforSinkTXslot}\right) = \frac{epoch\ \ size}{n\ \ downstream\ \ slots}$.

Chapter 7
Planning for Distributed Control Systems with Wired and Wireless Sensors

The planning algorithm that we describe next allows users to dimension the network and conclude whether desired latency bounds are met. Figure 7.6 shows a diagram of the planning approach proposed in this chapter. The chapter starts by introducing and discussing the user inputs. In Sect. 7.2, a general overview of the algorithm is done, and then, each step of the algorithm is described which allows to understand how it works and how to create a DCS with wireless sensor nodes and provides operation timing guarantees.

7.1 User Inputs

The algorithm allows users to dimension the network based on two alternatives. One alternative assumes the user provides a complete TDMA schedule and operation timing requirements. The algorithm checks if latencies are met and modifies the schedule (may even determine the need to divide the network) to meet the latency bounds.

The other alternative allows the user to specify only the minimum possible amount of requirements, and the algorithm creates the whole schedule and takes into consideration all constraints:

- *Network Layout*. The first network layout is completely defined by the user. It takes into account the physical position of the nodes, their relation (which nodes are leaf nodes and their parents), and a schedule to define how data and commands flow in the network.

  Appendix C shows a text-based example of how this could be specified.
- *Network Configuration and Data Forwarding Rule*. The network configuration is indicated by the user and takes into account the physical position of the nodes (which nodes are leaf nodes and their parents), and data forwarding rule indicates how the schedule must be defined to forward data messages from sensing

J. Cecílio and P. Furtado, *Wireless Sensors in Industrial Time-Critical Environments*, Computer Communications and Networks, DOI 10.1007/978-3-319-02889-7_7,

Fig. 7.1 Data forwarding rule – option 1

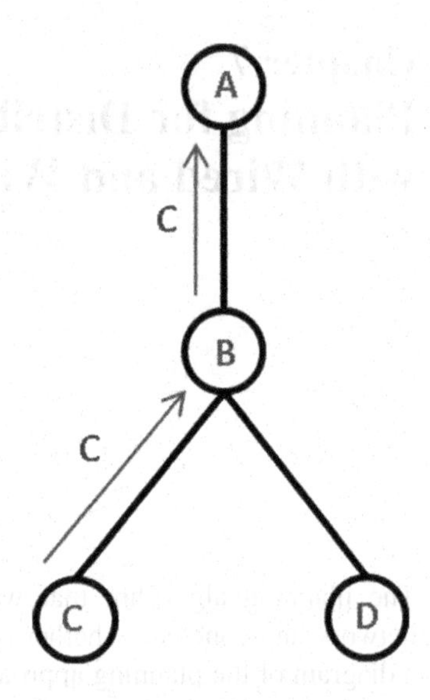

nodes to the sink node. The node slot positioning is directly dependable of the network configuration but should be optimized to reduce latencies or the number of radio wake-ups. The data forwarding rule can assume one of the following options:

- Each node sends data to its parent. Each parent receives data from a child and forwards immediately to its parents (Fig. 7.1).
- Each node sends data to its parent. Each parent collects data from all children and only forwards up after receiving from all children (Fig. 7.2).
- Each node sends data to its parent. Each parent collects data from all children, aggregates data messages from all children, and only forwards a merged message to its parents (Fig. 7.3).

  Appendix B shows one text-based example of how this alternative could be specified.

The algorithm assumes that users provide the information needed to use one of the above alternatives. It is also necessary for users to indicate some other parameters:

- $t_{Clock_{Sync}}$ – this is the clock synchronization interval, i.e., the time between clock synchronizations, which are necessary to keep the WSN node clocks synchronized, avoiding clock drifts between them.
- $\max(t_{CMD\text{Processing}})$ – this is a small time needed to parse and process a command in any WSN node. The user must specify a maximum bound for this time.
- $\max\left(t_{CL(i)_{\text{Processing}}}\right)$ – for each closed-loop operation, the user should specify the maximum required time taken to process, which corresponds to the computation time needed at the decision-maker node to take a decision, for commanding the required actuation.

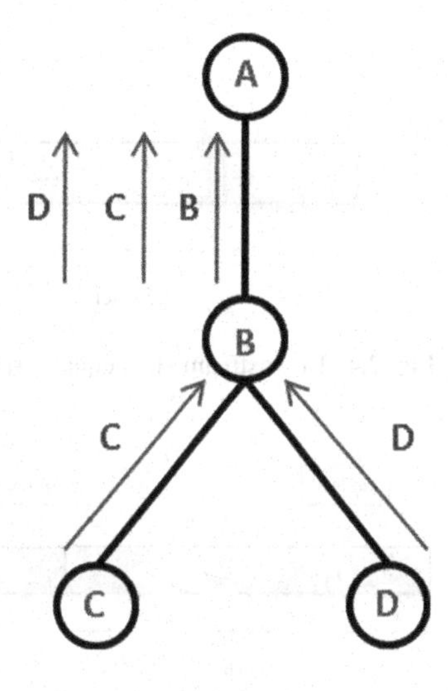

Fig. 7.2 Data forwarding rule – option 2

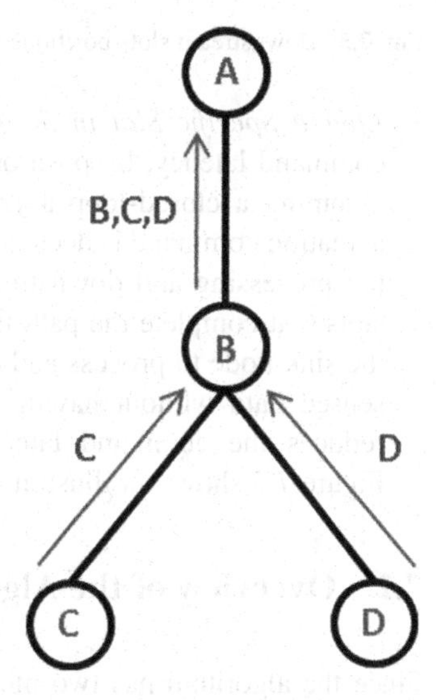

Fig. 7.3 Data forwarding rule – option 3

The values of $\max(t_{Serial})$ and $\max(t_{Middleware})$, defined in the previous chapter and used by the latency model, should be given by the user. They are previously obtained by distributed control system testing. A desired sampling rate should also be indicated.

Lastly, the algorithm needs to be configured concerning downstream slots positioning rule. The proposed algorithm supports two alternatives to position those:

- *Equally Spaced in the Epoch*. The downstream slots are positioned equally spaced in the epoch. Assuming that, due to latency requirements, two downstream slots are added by the algorithm, Fig. 7.4 shows how these slots are positioned according to this alternative.

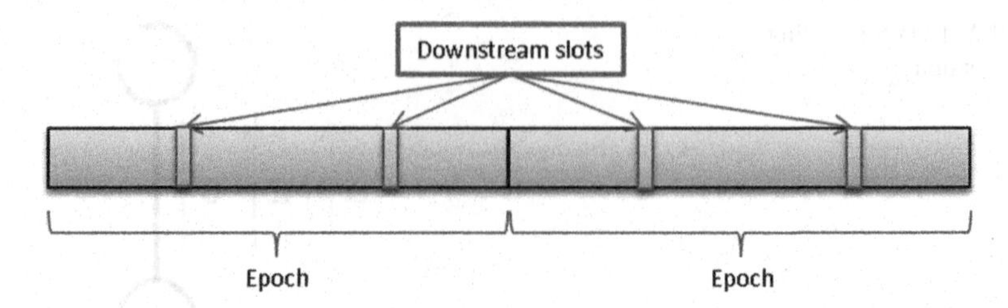

Fig. 7.4 Downstream slots equally spaced in the epoch

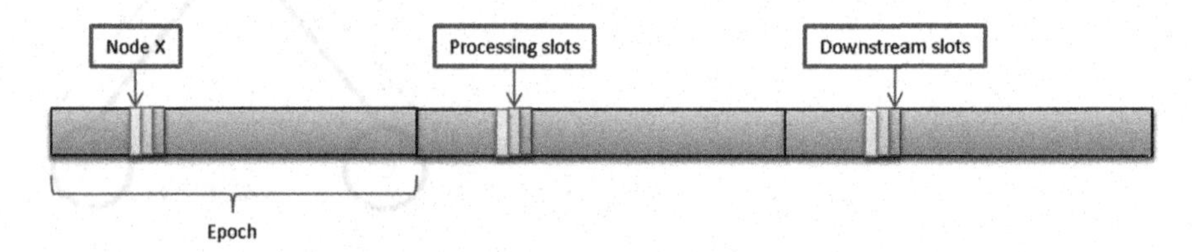

Fig. 7.5 Downstream slots positioned to optimize asynchronous closed-loop latency

- *After a Specific Slot in the Epoch.* It allows to optimize the processing and command latency, by positioning these slots after a specific slot. For instance, assuming a closed-loop asynchronous operation in the sink node, where the actuation command is decided based on sensed data coming from a sensor node, the processing and downstream slots should be positioned after the upstream slots that complete the path from the sensor node to the sink node. This allows the sink node to process and command an actuator immediately after receiving sensed data without having to wait more time for a downstream slot. This reduces the command latency and, consequently, the closed-loop latency. Figure 7.5 shows an illustration of this alternative.

7.2 Overview of the Algorithm

Since the algorithm has two modes of operation, if the user chooses to provide a network configuration plus forwarding rule as an input (see example in Appendix A), the algorithm starts by defining the upstream part of the TDMA schedule for all nodes, according to the forwarding rules (steps 1, 2, and 3). This results in a first TDMA schedule sketch, which has all upstream slots but still needs to be completed (Fig. 7.6).

Instead of this alternative, the user may have provided the network layout as an input (see example in Appendix B). In that case, the network layout given is already a TDMA schedule, and the algorithm checks and analyzes it to guarantee latencies.

After the first TDMA schedule and the current epoch size are determined, the algorithm analyzes latency requirements and determines how many downstream slots are needed (steps 5 and 6). The necessary number of slots is added to the

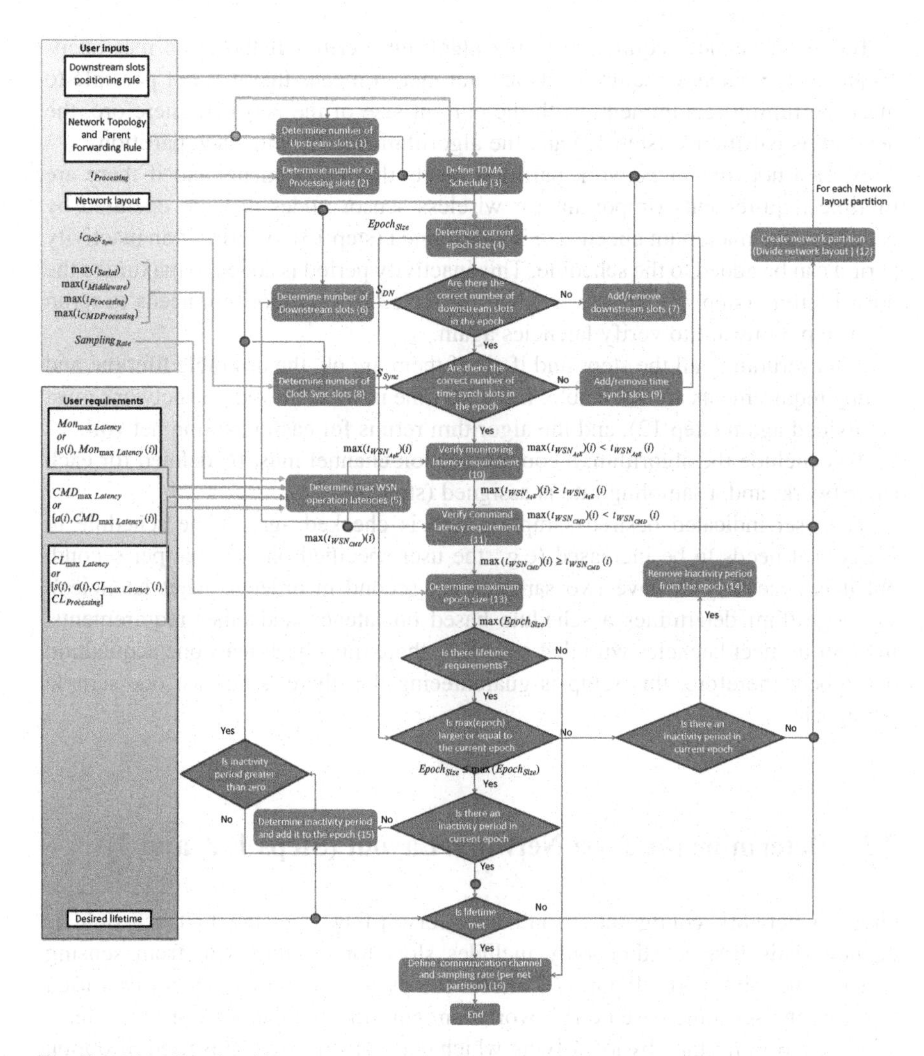

Fig. 7.6 Planning algorithm

TDMA schedule according to the processing and downstream slots positioning requirements (steps 7 and 3). The current epoch size is determined again. Next, based on $t_{Clock_{Sync}}$, the algorithm determines how many slots are needed for time synchronization and adds them to the schedule (steps 8 and 9). The new epoch size is determined, and if the user specified equally spaced downstream slots, the number of downstream slots is verified to check if they are enough for the current epoch size. If they are not enough, more downstream slots are added and a new schedule is recomputed.

Based on latency requirements, the algorithm verifies if those are met (steps 10 and 11). If latency requirements are not met, it means that it is not possible to meet the timing requirements with the current size of the network; therefore, the network is partitioned (step 12) and the algorithm restarts with each partition.

After a network or network partition meets the requirements and if there are lifetime requirements (important for wireless sensor nodes that are operated by batteries), the maximum epoch size is determined (step 13) to verify if an inactivity period can be added to the schedule. This inactivity period is added to maximize the node lifetime (step 15). However when we added it, the algorithm needs to rerun from step 3 onward to verify latencies again.

After rerunning all the steps and if all of them are ok, the desirable lifetime and timing requirements are attainable. If the lifetime is not achieved, the network must be divided again (step 12), and the algorithm reruns for each resulting network.

To conclude the algorithm, a communication channel must be defined for each subnetwork, and a sampling rate is assigned (step 16).

The user-indicated desired sampling rate is checked against the schedule to verify if it needs to be increased (e.g., the user specified one sample per second, but it is necessary to have two samples per second in order to meet latencies). The algorithm determines a schedule based on latency and other requirements. In order to meet latencies with that schedule, there must be at least one acquisition per epoch; therefore, this setup is guaranteeing that there is at least one sample per epoch.

7.3 Determine the First Network Layout (Steps 1, 2, and 3)

Given a network configuration and data forwarding rule, the first schedule is created. This first schedule only includes slots for sending data from sensing nodes to the sink node. Figure 7.7 shows the pseudo-code of the algorithm used to create the schedule based on network configuration and data forwarding rule.

The algorithm starts by identifying which data forwarding rule is used. If option 1, described in Sect. 7.1, is used, the algorithm will select node by node from the network configuration, determine the path to the sink, and allocate forwarding slots for the node and for each node between its position and the sink node.

If option 2 or 3 is used, the algorithm runs in a recursive way to allocate slots from the leaf nodes to the sink node. If option 2 is used, the algorithm allocates a slot per node plus one slot for each child node connected to it. In case of option 3, slots to forward data from the child nodes are not added. In this case, we assume that all data coming from child nodes can be merged with node data in a single message.

The slot assignment algorithm can also use one or more than one slot if retransmission is desired to increase reliability.

```
        if data forwarding rule == option 1 then
          for each node in the network configuration without
slot for its data transmission
              allocate a slot to it;
              look up for the path to the sink node;
              for each parent in the path to the sink node:
                allocate a slot to it;
        else if data forwarding rule == option 2 or option 3
then
          C = set of child nodes for node i;
          slotAssignment(C(sink));
        end if

        slotAssignment (C){
          for each node i in C
            if there are child nodes connected to it then
              slotAssignment (C (i));
            else
              allocate slot to node i
              if data forwarding rule == option 2 then
                allocate a slot for each child connected to
node i
        }
```

Fig. 7.7 Slot assignment algorithm – pseudo-code

7.4 Determine Current Epoch Size (Step 4)

The current epoch size is the number of slots in the current schedule. This schedule could be indicated by the user as an input or can result from applying the algorithm.

This step is recomputed several times during the algorithm flows because the current schedule is changing along the flow. For instance, if the user introduces a network configuration and data forwarding rule, the algorithm creates a whole schedule. In steps 1, 2, and 3, the algorithm creates a first schedule and determines the current epoch size. Based on this current epoch size, the algorithm determines how downstream slots are needed according to latency requirements and user inputs. After that, these downstream slots are added to the schedule, resulting in a different epoch size. So, step 4 is called again to determine the current epoch size.

7.5 Determine Maximum WSN Latencies (Step 5)

Once the schedule is defined (and consequently the epoch size), the algorithm will predict the WSN part of latency.

Assuming a wireless TDMA plus wired CSMA configuration and a maximum latency for monitoring messages ($Mon_{MaxLatency}$), based on Eq. 6.14, we determine the maximum admissible latency for the WSN subnetwork (Eq. 7.1).

$$\max\left(t_{WSN_{AqE}}\right) = Mon_{MaxLatency} - \left(\max\left(t_{Serial}\right) + \max\left(t_{Middleware}\right) + \max\left(t_{\text{Processing}}\right)\right) \quad (7.1)$$

Instead of a single maximum monitoring latency $Mon_{MaxLatency}$, the algorithm allows the user to specify pairs [$node(i)$, $Mon_{MaxLatency}(i)$]. In this case, Eq. 6.14 must be applied for each node, resulting in a maximum admissible latency per WSN node (Eq. 7.2).

$$\max\left(t_{WSN_{AqE}}\right)(i) = Mon_{MaxLatency}(i) - \left(\max\left(t_{Serial}\right) + \max\left(t_{Middleware}\right) + \max\left(t_{\text{Processing}}\right)\right) \quad (7.2)$$

Similar to the monitoring latency, users can define maximum latencies to deliver a command to a WSN node. Assuming that $CMD_{MaxLatency}$ is specified as maximum command latency, based on Eq. 6.18, $\max(t_{WSN_{CMD}})$ can be determined as

$$\max\left(t_{WSN_{CMD}}\right) = CMD_{MaxLatency} - \left(\max\left(t_{Middleware}\right) + \max\left(t_{Serial}\right) + \max\left(t_{CMD\text{Processing}}\right)\right) \quad (7.3)$$

If pairs of [$node(i)$, $CMD_{MaxLatency}(i)$] are given, the algorithm applies Eq. 7.3 per each pair, resulting in a set of maximum command latencies. Based on that set of latencies, the algorithm chooses the strictest latency, and dimensions the network to meet that latency (steps 6 and 7).

7.6 Determine the Number of Downstream Slots (Steps 6 and 7)

In Eq. 7.3 we determine $\max(t_{WSN_{CMD}})$. Through the discussion given in the previous chapter and Eq. 6.15, we can see that it is directly dependent on the number of the downstream slots per epoch.

Each epoch must accommodate slots for transmitting configuration and actuation commands. The minimum number of downstream slots (S_{DN}) that must exist can be calculated as follows:

$$S_{DN} = H \cdot s \quad (7.4)$$

where H represents the number of levels and represents the number of slots per level (one by default, plus one or more for enhanced reliability, to accommodate retransmissions). This is the case where the whole epoch has one downstream slot available for the sink node to forward a message downward.

The worst-case latency for this case is larger than an epoch size, as shown and discussed in Eqs. 6.14 and 6.15. Since epoch sizes may be reasonably large

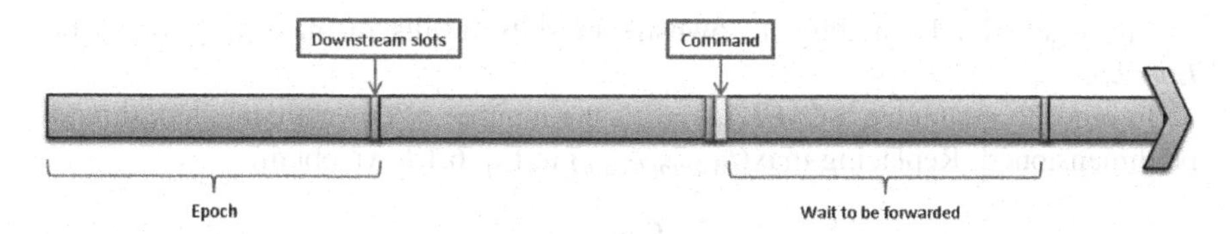

Fig. 7.8 Worst case: schedule with one downstream slot

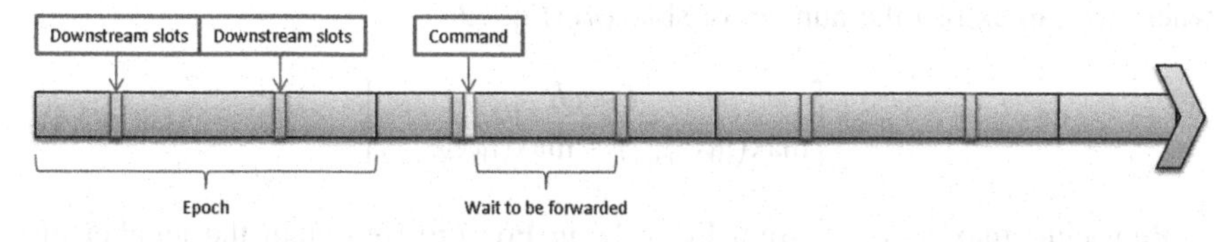

Fig. 7.9 Schedule with two downstream slots

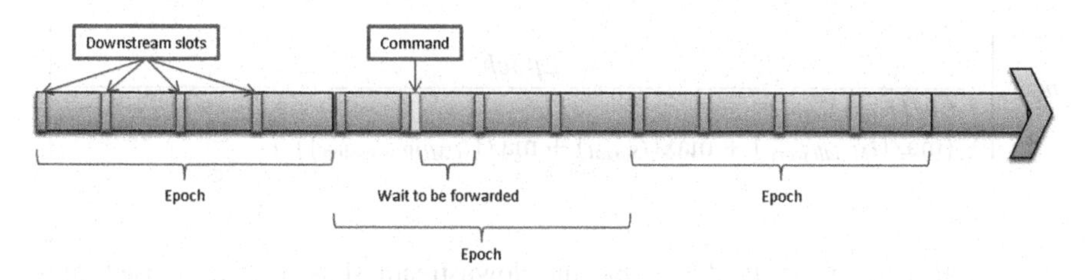

Fig. 7.10 Schedule with four downstream slots

(e.g., 1 s), this may result in an undesirable command latency, in particular it may not meet latency requirements.

The number of downstream slots can be increased to meet user command latency requirements. Figure 7.8 shows an example of a worst-case wait time for a command arriving at the sink and waiting for its downstream slot.

As shown in Fig. 7.8, if one downstream slot is provided per epoch, the command must wait, in the worst case, a full epoch to be transmitted. Additionally, in average a command will have to wait half the epoch size.

In order to shorten the max($Command_{Latency}$), there are two major alternatives: reducing the epoch length (we assume this is unwanted, since it was already determined to reflect a network with a certain number of nodes, energy, and latency requirements) and adding extra downstream slots.

Next we discuss the addition of more downstream slots to reduce $t_{WaitforTXslot}$ of Eq. 6.18. Those can be placed equally spaced in the epoch to minimize the maximum expected $t_{WaitforTXslot}$, given that number of downstream slots. As an example, Fig. 7.9 shows an epoch with two downstream slots. In this case, when a command arrives at the sink, it must wait a maximum time of *Epoch*/2.

In the next example (Fig. 7.10), the epoch includes four downstream slots which allows the maximum wait time to be reduced to *Epoch*/4.

More generically, adding n downstream slots results in $\max(t_{WaitforTXslot})$ of $Epoch/n$.

In order to guarantee a $CMD_{MaxLatency}$, the number of downstream slots should be dimensioned. Replacing $\max(t_{WaitforTXslot})$ in Eq. 6.18, we obtain

$$\max\left(t_{WSN_{CMD}}\right) = \frac{Epoch}{n} + \max\left(t_{WSN_{Down}}\right) \tag{7.5}$$

where we can extract the number of slots (n) (Eq. 7.6).

$$n = \left\lceil \frac{Epoch}{\max\left(t_{WSN_{CMD}}\right) - \max\left(t_{WSN_{Down}}\right)} \right\rceil \tag{7.6}$$

Replacing $\max\left(t_{WSN_{CMD}}\right)$ from Eq. 6.18 in Eq. 7.6, we obtain the number of downstream slots in function of $CMD_{MaxLatency}$ (Eq. 7.7).

$$n = \left\lceil \frac{Epoch}{\left(\begin{array}{l} CMD_{MaxLatency} - \\ \left(\max(t_{Middleware}) + \max(t_{Serial}) + \max\left(t_{CMD\text{Processing}}\right)\right) \end{array} \right) - \max\left(t_{WSN_{Down}}\right)} \right\rceil \tag{7.7}$$

It is also necessary to dimension the downstream slots to meet closed-loop operation latency requirements. If a synchronous closed-loop operation is defined, $CL_{MaxLatency}$ corresponds to $CMD_{MaxLatency}$ latency, because in this case we consider only the latency from the closed-loop supervisor to the actuator. Therefore, the number of downstream slots is determined by Eq. 7.7.

If an asynchronous closed-loop operation is considered, $CL_{MaxLatency}$ includes monitoring and command parts (Eq. 7.8).

$$\begin{aligned} \max\left(CL_{Latency_{Async}}\right) = {} & \max\left(t_{WSN_{AqE}}\right) + \max(t_{Serial}) + \max(t_{Middleware}) + \max\left(t_{\text{Processing}}\right) \\ & + \max(t_{Middleware}) + \max(t_{Serial}) + \max\left(t_{WSN_{CMD}}\right) + \max\left(t_{CMD\text{Processing}}\right) \end{aligned} \tag{7.8}$$

Replacing Eq. 7.5 in Eq. 7.8 and applying mathematical operations, we obtain the number of downstream slots to meet asynchronous closed-loop latencies (Eq. 7.9).

$$n = \left\lceil \frac{Epoch}{\begin{pmatrix} \max\left(\mathrm{CL}_{Latency_{Async}}\right) \\ -\begin{pmatrix} \max\left(t_{WSN_{AqE}}\right) + \max(t_{Serial}) + \max(t_{Middleware}) + \max\left(t_{\mathrm{Processing}}\right) + \\ \max(t_{Middleware}) + \max(t_{Serial}) + \max\left(t_{CMD\mathrm{Processing}}\right) \end{pmatrix} \\ -\max(t_{WSN_{Down}}) \end{pmatrix}} \right\rceil \tag{7.9}$$

If we assume that the serial and middleware latencies for upstream and downstream are equal, Eq. 7.9 may be simplified and results in Eq. 7.10.

$$n = \left\lceil \frac{Epoch}{\left(\max\left(CL_{Latency_{Async}}\right) - \begin{pmatrix} \max\left(t_{WSN_{AqE}}\right) \\ +2\cdot\max(t_{Serial}) \\ +2\cdot\max(t_{Middleware}) \\ +\max\left(t_{\mathrm{Processing}}\right) \\ +\max\left(t_{CMD\mathrm{Processing}}\right) \end{pmatrix} - \max(t_{WSN_{Down}}) \right)} \right\rceil \tag{7.10}$$

After determining the number of downstream slots needed to meet command latency or closed-loop latency requirements, the algorithm adds them to the schedule and recomputes the current epoch size (step 4).

7.7 Number of Clock Synchronization Slots (Steps 8 and 9)

When a TDMA protocol is assumed, the algorithm needs to add specific slots for clock synchronization ($N_{Slots_{Sync}}$). $N_{Slots_{Sync}}$ is determined based on the and the parameter clock interval ($t_{Clock_{Sync}}$), which is indicated by the user. It is determined as follows:

$$N_{Slots_{Sync}} = \left\lceil \frac{Epoch}{t_{Clock_{Sync}}} \right\rceil \tag{7.11}$$

After determining $N_{Slots_{Sync}}$, the algorithm adds them to the schedule and recomputes the current epoch size (step 4).

7.8 Verify if Latency Requirements Are Met with the Current Epoch, Network Layout, and Schedule (Steps 10 and 11)

Based on Eqs. 6.2 and 6.6, the algorithm determines the latency required to transmit a data message from a WSN leaf node to the sink node ($t_{WSN_{AqE}}$) and the latency to send a command from the sink node to the WSN leaf node ($t_{WSN_{CMD}}$).

After determining these two latencies, the algorithm compares the values with the latency determined through Eqs. 6.15 and 6.19, which results from the user requirements, and concludes if all latencies are met.

If any of the latencies are not met, the algorithm needs to partition the network to find a schedule and an epoch size for each new subnetwork that meets the requirements (step 12).

7.9 Network Partitioning (Step 12)

When the algorithm detects that a network must be divided to meet user requirements, step 12 is called.

Assuming that a user gives a network configuration and a parent forwarding rule, the algorithm divides the initial network configuration in two parts. This division is done automatically, if the initial network configuration has the same configuration for all branches. Otherwise, the user is requested to split the network and restart the algorithm with the new configuration.

On the other hand, if a network layout is given, the algorithm divides the network and the schedule into two parts. The downstream, processing, and clock sync slots are copied for both parts.

7.10 Determine the Maximum Epoch Size (Steps 13 and 14)

The maximum epoch size is defined as the epoch size which is able to guarantee desired latencies and conforms to the sampling rate. This can be determined as

$$\max(Epoch_{Size}) = \min\left\{Sampling_{Rate}, \max(Epoch_{Size})_{Latency}\right\} \tag{7.12}$$

where $\max(Epoch_{Size})_{Latency}$ is determined according to the following configurations:

- If acquisition instant is not synchronized with sending instant and we do not consider the event start instant, $\max\left(t_{WSN_{AqE}}\right)$ corresponds to the sum of the time waiting for the transmission slot plus the time taken for the data to travel from source node to the sink node. In this case, $\max(Epoch_{Size})_{Latency}$ is defined by

$$\max(Epoch_{Size})_{Latency} = \max\left(t_{WSN_{AqE}}\right) - \left(\max\left(t_{Aq}\right) + \max\left(t_{WSN_{UP}}\right)\right) \quad (7.13)$$

- If acquisition instant is synchronized with sending instant and we are considering the event occurrence instant, $\max\left(t_{WSN_{AqE}}\right)$ corresponds to the time for event detection and its transmission from source node to the sink node. In this case, $\max(Epoch_{Size})_{Latency}$ assumes the same value of the previous case, as defined by Eq. 7.13.
- If acquisition instant is not synchronized with sending instant and we are considering the event occurrence instant, $\max\left(t_{WSN_{AqE}}\right)$ corresponds to the time for event detection, plus the time waiting for the transmission instant, plus the travel time from source node to sink node. In this case, $\max(Epoch_{Size})_{Latency}$ is defined as

$$\max(Epoch_{Size})_{Latency} = \frac{\max\left(t_{WSN_{AqE}}\right) - \left(\max\left(t_{Aq}\right) + \max\left(t_{WSN_{UP}}\right)\right)}{2} \quad (7.14)$$

In these equations, the maximum WSN latency ($\max(t_{WSN_{UP}})$), which was defined in the latency model description (previous chapter), corresponds to the time taken to transmit a message from a leaf node to the sink node using the current network layout. It depends on the maximum number of levels included in the monitoring operation. t_{Aq} corresponds to the acquisition latency.

If the $\max(Epoch_{Size})_{Latency}$ is smaller than the current epoch size (from step 4), then either cut the inactivity period of the current schedule (step 14) or otherwise divide the network (step 12) because it is not possible to meet latency requirements or sampling rate with the current size of the network. The sampling rate parameter for each new subnetwork will be defined as the user-defined sampling rate.

7.11 Inactivity Period (Step 15)

The nodes of the WSNs may or may not be battery operated. If they are battery operated, the user may have defined lifetime requirements or he or she may have specified that he wants to maximize lifetime. In order to meet lifetime requirements, it may be necessary to add an inactivity period to the epoch, during which nodes have low consumption, because they turn their radio off.

Latency and sampling rate specifications may restrain the inactivity period that would be necessary to guarantee the desired lifetime. In that case, the algorithm can divide the network into two subnetworks and rerun over each, trying to accommodate both the latency and lifetime requirements.

The inactivity period can be determined as follows:

$$Inactivity_{Period} = \min\{Inactivity_{Period}(\max(Epoch_{Size})), Inactivity_{Period}(Lifetime)\} \quad (7.15)$$

where the inactivity period due to the epoch size ($Inactivity_{Period}(\max(Epoch_{Size}))$) is determined by Eq. 7.16.

$$Inactivity_{Period}(\max(Epoch_{Size})) = \max(Epoch_{Size}) - Epoch_{Size} \quad (7.16)$$

The inactivity period required by lifetime requirements is defined in Sect. 7.13. After determining this $Inactivity_{Period}$ quantity, it is added to the schedule, and the algorithm restarts again to verify if all constraints are achieved. This is required because the addition of the inactivity period may have consequences concerning command latencies and the synchronization slots that need to be recalculated.

7.12 Network Communication Channel and Sampling Rate (Step 16)

Since the initial network may need to be divided into several smaller networks, different communication channels should be defined for each resulting subnetwork to avoid communication interferences between different subnetworks.

Lastly, the sampling rate is defined as the epoch size or a subsampling alternative if the user wishes:

$$Sampling_{rate} = Epoch_{Size} \quad (7.17)$$

If users want to have subsampling (multiple samples per sampling period), the sampling rate would be

$$Sampling_{rate} = \frac{Epoch_{Size}}{n} \quad (7.18)$$

where n represents the number of samples per period. Multiple samples per period allow, for instance, to apply smoothing operations in order to remove noise.

7.13 Lifetime Prediction

Energy consumption is important in networked embedded systems for a number of reasons. For battery-powered nodes, energy consumption determines their lifetime.

The radio transceiver is typically the most power-consuming component. In a TDMA protocol, all network nodes are synchronized to a common clock. Nodes wake up on dedicated time slots at which they are ready to either receive or transmit data.

The node energy consumption can be estimated as [1, 2]

$$E = \left(I_{on} \cdot Radio_{DutyCycle} + \left(1 - Radio_{DutyCycle}\right) \cdot I_{off}\right) \cdot T \cdot V_{Bat} \tag{7.19}$$

where the radio duty cycle is measured as

$$Radio_{Dutycyle} = \frac{t_{Active(Listen+Transmit)}}{Epoch} \tag{7.20}$$

The currents I_{on} and I_{off} represent the amount of consumed current in each state of the radio (*on* and *off*). V_{Bat} is the power supply voltage to power up the node and T represents the amount of time where E is measured.

So, the node lifetime (T_{Total}) can be extracted from Eq. 7.19, where E represents the total charged battery capacity.

$$T_{Total} = \frac{E}{\left(I_{on} \cdot Radio_{DutyCycle} + \left(1 - Radio_{DutyCycle}\right) \cdot I_{off}\right) \cdot V_{Bat}} = Node_{Lifetime} \tag{7.21}$$

Consequently, network lifetime is defined as the lifetime of the node which discharges its own battery first.

In the previous section we discussed how to plan a network to meet latency and lifetime requirements. To optimize the TDMA schedule and to provide desirable lifetime, an inactivity period must be added to the schedule. Through Eq. 7.15, we define this inactivity period as

$$Inactivity_{Period} = \min\{Inactivity_{Period}(Epoch_{Size}), Inactivity_{Period}(Lifetime)\} \tag{7.22}$$

Therefore, the $Inactivity_{Period}(Lifetime)$ can be calculated through Eq. 7.23, replacing the $Radio_{Dutycyle}$ defined in Eq. 7.20 and solving it.

$$Inactivity_{Period}(Lifetime) = \left(\frac{t_{Active(Listen+transmit)} \cdot T_{Total} \cdot \left(I_{off} \cdot V_{Bat} - I_{on}\right)}{I_{off} \cdot T_{Total} \cdot V_{Bat} - E}\right) - C\_Epoch_{Size} \tag{7.23}$$

where C_Epoch_{Size} represents the current epoch size resulting from step 4 of the algorithm.

7.14 Slot Size Considerations

The slot size should be as small as possible to reduce the epoch size and consequently the end-to-end latency. To determine the slot time, the following times must be taken into account:

- Time to transfer a message from the MAC layers data FIFO buffer to buffer of radio transceiver (t_{ts})
- Time to transmit a message (t_{xm})
- Time a receiver needs to process the message and initiate the transmission of an acknowledgment message (t_{pm})
- Time to transmit an acknowledgment (t_{xa})
- Time to transfer and process the acknowledgment from the radio transceiver and to perform the associated actions for received/missed acknowledgment (t_{pa})

Also, a small guardian time is required at the beginning and end of each slot to compensate for clock drifts between nodes (t_g). Thus, the minimum size of a transmission slot is given as

$$T_{st} = t_{ts} + t_{xm} + t_{pm} + t_{xa} + t_{pa} + t_g \tag{7.24}$$

In our test bed, the slot size was 10 ms, which allows starting the communication, sending a data packet with 128 bytes of maximum size, and receiving the acknowledgment.

References

1. Polastre J, Szewczyk R, Culler D (2005) Telos: enabling ultra-low power wireless research. In: Proceedings of the IPSN 2005 fourth international symposium on information processing in sensor networks 2005, vol 00, No. C. IEEE Press, Piscataway, NJ, USA, pp 364–369. ISBN:0-7803-9202-7
2. Nguyen HA, Forster A, Puccinelli D, Giordano S (2011) Sensor node lifetime: an experimental study. In: Proceedings of the 2011 I.E. international conference on pervasive computing and communications workshops PERCOM workshops, IEEE, pp 202–207

Chapter 8
Performance and Debugging

Operations performance monitoring is important in contexts with timing constrains. For instance, in the previous chapter, we propose and discuss an algorithm to plan for timing guarantees in distributed control systems with heterogeneous components. In this chapter we define measures and metrics for surveillance of expectable time bounds and an approach for performance monitoring bounds on these metrics.

This surveillance can be used in any distributed system to verify performance compliance. Assuming that we have monitoring or closed-loop tasks with timing requirements, this allows users to constantly monitor timing conformity.

In the context of networked control systems with heterogeneous components and non-real-time parts, where latencies were planned, this allows the system to monitor and determine the conformity with timing requirements.

We define measures and metrics which create an important basis for reporting the performance to users and for helping them to adjust deployment factors. Those sets of measures and metrics are also used for debugging, using tools and mechanisms to explore and report problems. We also propose an approach to monitor the operation timings.

The time bounds and guarantees must be based on well-defined measures and metrics. In Sects. 8.1 and 8.2, we discuss these measures and metrics. Section 8.3 discusses metric information for analysis. We will discuss time bounds setting and time bounds metrics and how message loss information is collected.

The measures can be taken per message or statistically per time periods. We describe how both alternatives are used in the approach. Measures can also be classified. Each message is classified according to each time bound as in-time, out-of-time, waiting-for, or lost message.

Section 8.4 describes the addition of debugging modules to the DCS architecture, and Sects. 8.5 and 8.6 describe debugging node component and operation performance monitor component. It is described how the performance information is collected and processed. An example of operation performance monitor UI is presented, which allows users to evaluate the performance.

J. Cecílio and P. Furtado, *Wireless Sensors in Industrial Time-Critical Environments*, Computer Communications and Networks, DOI 10.1007/978-3-319-02889-7_8,

8.1 Measures

Operation timing issues in terms of monitor and closed-loop control can be controlled with the help of two measures, which we denote as *latency* and *delay of periodic events*.

8.1.1 Latency

Latency consists of the time required to travel between a source and a destination. Sources and destinations may be any site in the distributed system. For instance, the latency can be measured from a WSN leaf node to a sink node, or from a WSN sensing node to a computer, control station or backend application. It may account for the sum of all components, including queuing and the propagation times, and it is additive – the latency between two points is the sum of the latencies in the path going through all intermediate points that may be considered between the two points. Figure 8.1a shows an example of how latency is measured when a leaf node transmits its data to a sink node in a 3–3 tree topology.

A represents the instant when message transmission starts. The transmission takes several milliseconds and the message is received by an intermediate node at instant *B*. The intermediate node saves the message in a queue until it gets its slot transmission time. When the slot is acquired (instant *C*), the message is transmitted to the next upper level and reaches the sink node at instant *D*. In the example 6a, the latency is given by the sum of all latency parts. (($B - A$) gives the latency from leaf node to the intermediate node, ($C - B$) gives the queue latency, and ($D - C$) gives the latency from intermediate node to sink node).

Figure 8.1b shows the latency from the same node considered in Fig. 8.1a, but in this case, the destination point is a control station. It includes the previous latency and adds the latency from the sink node to the control station. Latency from sink node to

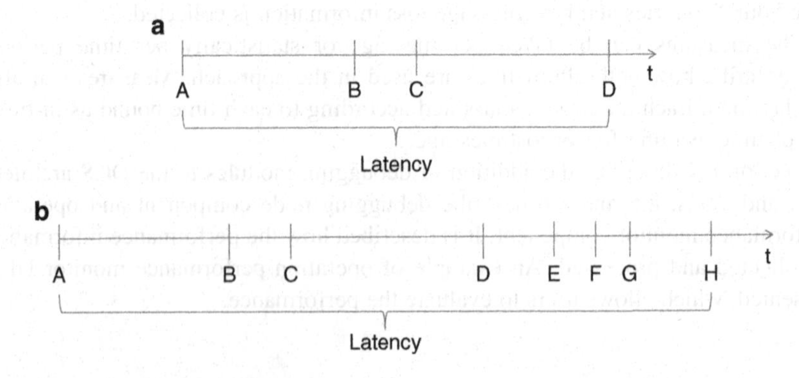

Fig. 8.1 Latency diagram (**a**) latency at sink node, (**b**) latency at control station

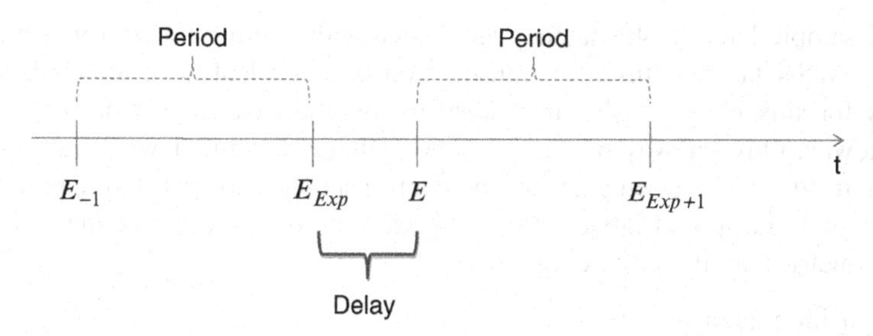

Fig. 8.2 Delay illustration

the control station is given by the sink node to gateway latency ($E - D$), plus gateway processing time ($F - E$), plus LAN transmission ($G - F$), plus control station processing time ($H - G$).

8.1.2 Delay of Periodic Events

Given a specific periodic event, such as monitoring or closed-loop control, the delay will be the extra time taken to receive a message with respect to the predefined periodic reception instant. Figure 8.2 shows how the delay is measured.

The instant E_{-1} represents the instant of the last occurrence of a specific periodic event. It is expected to receive that event with a specific cadence (period). E_{Exp} represents the instant of expected reception of the event. But the event may arrive delayed (instant E). So, the delay is the time elapsed between E_{Exp} and E instants. Each period time interval is measured from the last reception instant to the next reception instant.

8.2 Metrics

Given the above time measures, we define metrics for sensing and control. The metrics allow us to quantify timing behavior of monitoring and closed loops.

8.2.1 Monitoring Latencies

Monitoring latency is the time taken to deliver a value from sensing node to the control station, for display or alarm computation. If a value determines an alarm, it will be the time taken since the value (event) happens to the instant when the alarm is seen in the control station console. Since the value must traverse many nodes and parts of the system, this latency can be decomposed into latencies for each part of the path. It is useful to collect the various parts of the latency – acquisition latency (sensor sample latency plus latency associated with waiting for data transmission instant), WSN latency (time for transmission between leaf node and sink node), latency for sink gateway (the time taken for the message to go from the sink to the gateway, plus gateway processing time), latency for middleware transmission (e.g., transmission between gateway and control station), control station processing latency, and end-to-end latency (leaf node to control station). All the following latency metrics are therefore considered:

- Acquisition latency
- WSN latency
- WSN to gateway interface latency
- Middleware latency
- Control station processing latency
- End-to-end latency

8.2.2 Monitoring Delays

The delay measure was already defined as the amount of extra time from the moment when a periodic operation was expected to receive some data to the instant when it actually received. When users create a monitoring task, they must specify a sensing rate. The control station expects to receive the data at that rate, but delays may happen in the way to the control station, therefore delays are recorded.

8.2.3 Closed-Loop Latency for Asynchronous or Event-Based Closed Loops

In this case, the closed-loop latency is the time taken from sensing node to actuator node, passing through the supervision control logic. It will be the time taken since the value (event) happens at a sensing node to the instant when the action is performed at the actuator node. Since the value must cross several parts of the system, this latency can be decomposed into latencies for each part of the path: upstream part (from sensing node to the control station) and downstream part (from control station to the actuator). The first part (upstream) is equivalent to monitoring latency and can be subdivided in the same subparts. The second part (downstream) corresponds to the path used by a command to reach an actuator. The following latency metrics should be considered to determine the closed-loop latency:

- Acquisition latency
- WSN upstream latency
- WSN to gateway interface latency
- Middleware latency
- Control station processing latency

- Middleware latency
- Gateway to WSN interface latency
- WSN downstream latency
- Actuator processing latency
- End-to-end latency

8.2.4 Closed-Loop Latency for Synchronous or Periodic Closed Loops

Synchronous or periodic closed loops can be associated with two latencies: monitoring latency and actuation latency. The monitoring latency can be defined as the time taken from sensing node to the supervision control logic (monitoring latency). The actuation latency corresponds to the time taken to reach an actuator. We also define an end-to-end latency as the time from the instant when a specific value is sensed and the moment when an actuation is done which incorporates a decision based on that value.

The following latency metrics should be considered to determine the closed-loop latency for synchronous or periodic closed loops:

- Acquisition latency
- WSN upstream latency
- WSN to gateway interface latency
- Middleware latency
- Wait for the actuation instant latency
- Control station processing latency
- Middleware latency
- Gateway to WSN interface latency
- WSN downstream latency
- Actuator processing latency
- End-to-end latency

8.2.5 Closed-Loop Delays

In synchronous closed-loop operations, actuation is expected within a specific period. However, operation delays may occur in the control station and/or command transmission. The closed-loop delay is the excess time.

In asynchronous closed loops, there can be monitoring delays. This means that a sample expected every x time units may be delayed.

In summary, the proposed measures and metrics include:

- Delays:
  - Monitoring delay
  - Synchronous closed-loop actuation delay

- Latencies:
  - Monitoring latencies:
    - All – end-to-end latency
    - Acquisition latency
    - WSN latency
    - WSN to gateway interface latency
    - Middleware latency
    - Control station processing latency
  - Closed-loop latencies:
    - All – CL end-to-end latency
    - Acquisition latency
    - WSN upstream latency
    - WSN to gateway interface latency
    - Middleware latency
    - Wait for the actuation instant latency (synchronous)
    - Control station processing latency
    - Middleware latency
    - Gateway to WSN interface latency
    - WSN downstream latency
    - Actuator processing latency

8.3 Metric Information for Analysis

For each of the previous metrics, we can have per-message values, per-time interval statistics, as well as per-time interval bounds statistics. In this section we define bounds and describe how each message is classified according to each bound. Then we describe how to count message and packet losses.

8.3.1 *Bounds: Waiting, In-Time, Out-of-Time, Lost*

Bounds over latency and delay measures allow users to configure bounds violation alarms and to keep information on how often the bounds are broken. A bound may be defined as a threshold over some measure. It specifies an acceptable limit for that measure. We classify the events with respect to that bound as:

Waiting: the process is waiting to receive the event; its final status with respect to the bound is yet undefined.
In-Time: the event arrived, and the time measure is within the specified bound.
Out-of-Time: the event arrived, and the time measure is out of the bound.
Lost: the event did not arrive, and the timeout has expired.

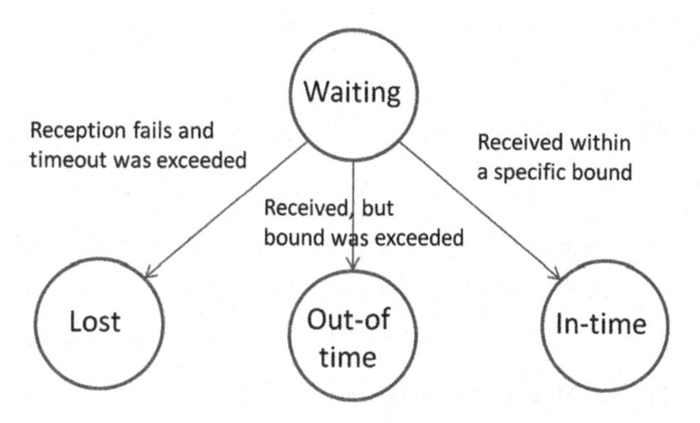

Fig. 8.3 Event-state diagram

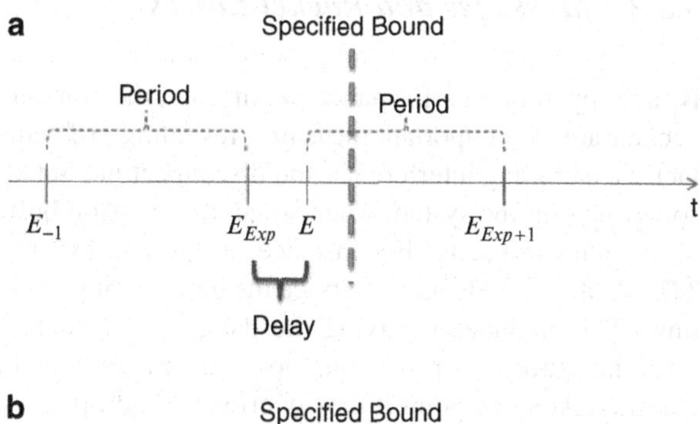

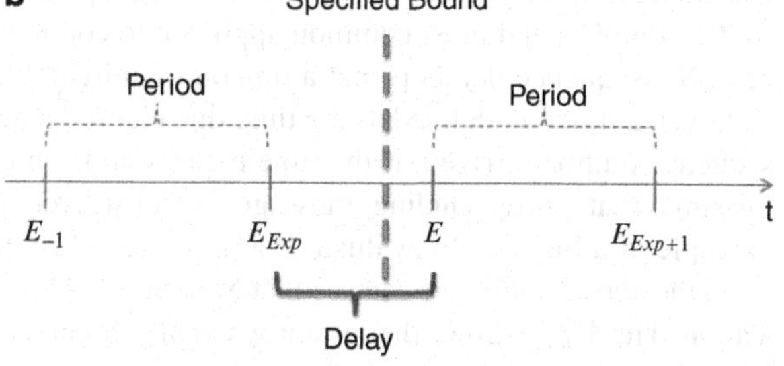

Fig. 8.4 Bounded event classification (**a**) in-time, (**b**) out-of-time

Figure 8.3 shows a state diagram with message classification. Each event is classified as "waiting" until reception or lost if timeout was exceeded. When an event is received within a specific bound, it is classified as "in-time." If an event was received but the bound was exceeded, it is classified as "out-of-time." Lastly, if an event is expected but is not received and the timeout has elapsed, it is classified as "lost."

Figure 8.4 shows an example of bound specification and corresponding event classification (in-time and out-of-time). Figure 8.4a shows an event (E) that arrives at destination with some delay but within a specified bound. In this case, the event is classified as in-time. Figure 8.4b shows an example where an event is classified as out-of-time. In this case the delay is greater than the specified bound.

It should be possible to specify delay bounds, latency bounds, or both.

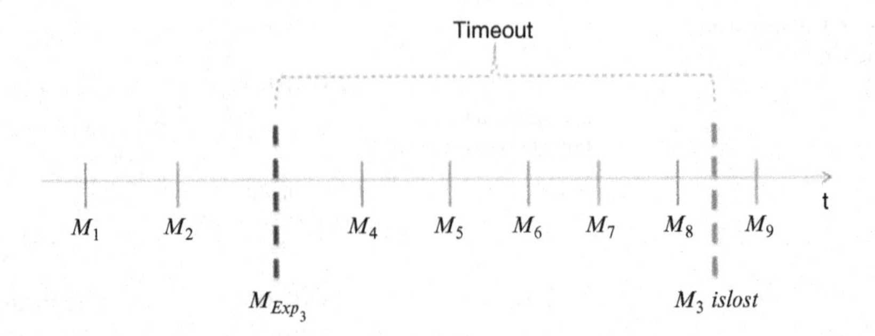

Fig. 8.5 Message lost evaluation

8.3.2 Messages and Packet Losses

Besides operation performance monitoring, the number or ratio of lost messages or packets are also important measures, revealing problems that may be due to several factors, including interference and disconnection, but also messages being dropped somewhere in the system when some transmission buffer or queue fills up for lack of servicing capacity. For instance, in the context of a preplanned network with TDMA, a sink node has to service the data coming from one node in a single slot of time. This includes receiving the data, sending it to the gateway through some serial interface, and processing downstream command messages coming from the gateway. At some point it may overload and drop messages.

The simplest and most common approach to count end-to-end message losses is based on sequence numbers and a timeout (configurable).

Given a timeout, defined as the time that elapsed since a message with a specific sequence number arrived, if the time expires and some lower sequence number is missing, that corresponding message is considered lost. Figure 8.5 shows an example of a timeline to evaluate if a message is lost or not.

In the above figure, we can see that Message 3 (M_3) does not arrive at the control station. The M_{Exp_3} shows the instant when M_3 should be received. M_3 is considered lost when the timeout is exceeded.

In closed-loop control scenarios, actuation command losses are accounted for by means of timeout in an acknowledgment message. Each actuation command has an acknowledgment message associated with it. A command is marked as lost when the acknowledgment is not received by the command issuer in a certain timeout after the command message was sent.

Network protocol-level count of packet losses can also be used to analyze losses.

8.3.3 Statistics: Avg, Stdev, Max, Min, Percentile

Statistic information can be computed for each part of the system (e.g., WSN part, gateway interface, gateway processing), and it will be important to diagnose timing problems and where they come from. The following types of statistical information are included:

Maximum: This is useful to detect the worst-case time measure, one that should not be over a limit.

Percentile (e.g., P99, P95, P90): It is useful to characterize the measure under evaluation, while removing outlier events (e.g., large delays existing less than 0.01 % of the cases).

In systems with strict timing requirements, maximum and percentile measures are the most important.

Averages and Standard Deviation: Which is another common metric used to evaluate the stability of time measures.

Minimum: It is a less important metric in our context but provides the best-case time measure.

These are important metric data items for reporting the performance to users and for helping users adjust their deployment factors. Another important aspect is that this statistic information (in short avg, stdev, max, percentile, min) can be taken for different periods of time, for each node or part of the system, and it is important to keep those with different granularities, so that it becomes possible to arrive at the trouble spots. Finally, it is useful to correlate those with operating system and networks performance information, to detect the culprit of the problems.

8.4 Addition of Debugging Modules to a Distributed Control System Architecture

In the previous sections we discussed measures and metrics useful to evaluate operation performance. In this section we will discuss how to add debugging modules to DCS architectures.

A DCS can be divided into two main parts: nodes and remote control components (control stations). To add debugging functionalities, we need to add a debugging module (DM) to each node and a Performance Monitor Module (PMM) to each control station, as illustrated in Fig. 8.6.

The debugging module collects information from operation execution in a node and then formats and forwards information to the Performance Monitor Module. The Performance Monitor gathers the status information coming from nodes, stores it in a database, and processes it according to bounds defined by the user.

In the next sections, we describe how the debugging module and Performance Monitor Module work.

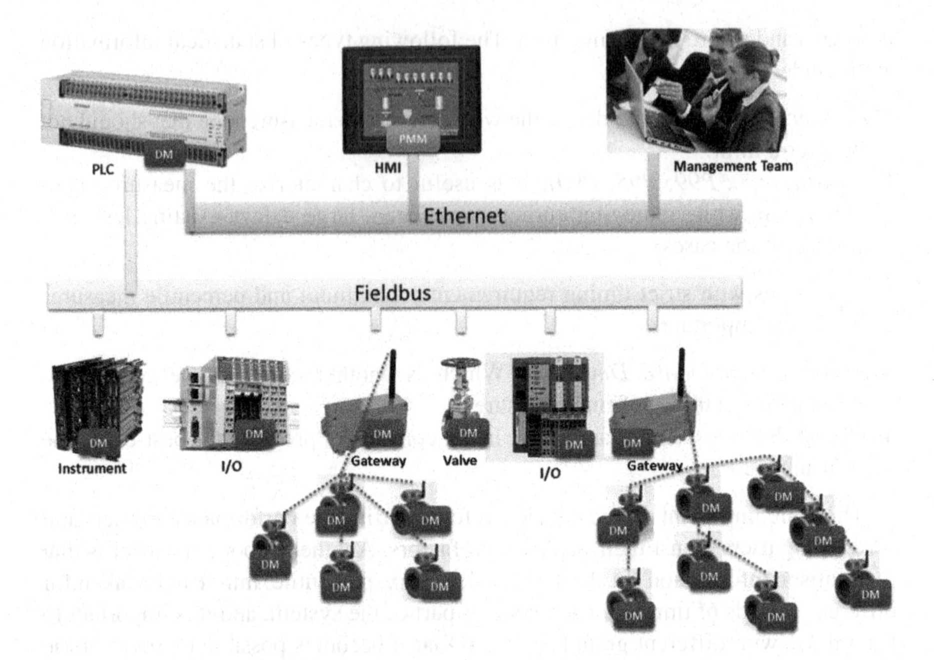

Fig. 8.6 DCS architecture with debugging modules

8.5 The Debugging Module

The debugging module (DM) stores all information concerning node operation (e.g., execution times, battery level) and messages (e.g., messages received, messages transmitted, transmission fails, transmission latencies). This information is stored inside the node. It can be stored either in main memory, flash memory, or other storage device.

DM is an optional module that can be activated or deactivated. It generates a debugging report, either by request or periodically, with a configurable period.

DM has two modes of operation:

- Network Debugging. The DM runs in all nodes and keeps the header information of messages (source ID, destination ID, Msg Type, Ctrl Type and Sequence Number), where it adds timestamps corresponding to arrive and departure instants. After, this information is sent periodically or by request to the Performance Monitor (described in the next section), which is able to calculate metrics. This operation mode may be deactivated in constrained devices, because it consumes resources such as memory and processing time.
- High-Level Operation Debugging. Instead of collecting, storing, and sending all information to the Performance Monitor, the DM can be configured to only add specific timestamps to messages along the path to the control station.

Assuming a monitoring operation in a distributed control system with WSN subnetworks, where data messages are sent through a gateway, the DM can be

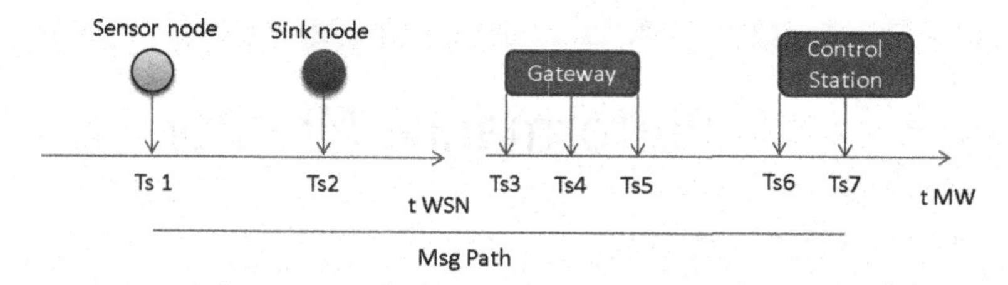

Fig. 8.7 Message path example

configured to add timestamps in the source node, sink node, gateway, and control station. Figure 8.7 illustrates nodes, gateways, and a control station in that context.

The approach assumes that WSN nodes are clock synchronized. However, they may not be synchronized with the rest of the distributed control system. Gateways, computers, and control stations are also assumed clock synchronized (e.g., the NTP protocol can be used).

In Fig. 8.7, the DM starts by adding a generation timestamp (source timestamp) in the sensor node (*Ts1*). When this message is received by the sink node, it adds a new timestamp (*Ts2*) and indicates to the gateway that a message is available to be written in the serial interface. Upon receiving this indication, the gateway keeps a timestamp that will be added to the message (*Ts3*), and the serial transmission starts. After concluding the serial transmission, the gateway takes note of the current timestamp (*Ts4*) and adds *Ts3* and *Ts4* to the message.

Upon concluding this process and after applying any necessary processing to the message, the gateway adds another timestamp (*Ts5*) and transmits it to the control station. When the message is received by the control station, it adds a timestamp (*Ts6*), processes the message, and adds a new timestamp (*TS7*), which indicates that the instant of message processing at the control station was concluded. After that, at the control station, the Performance Monitor Module (described in the next section) receives the message, and, based on the timestamps that come in the message, it is able to calculate metrics.

If there is only one computer node and control station, there will only be *Ts1*, *Ts6*, and *Ts7*.

8.6 The Performance Monitor Module and UI

In this section we describe the Performance Monitor Module (PMM), which debugs operations performance in the heterogeneous distributed system. The PMM stores events (data messages, debug messages), latencies, and delays into a database. It collects all events when they arrive, computes metric values, classifies events with respect to bounds, and stores the information in the database. Bounds should be configured for the relevant metrics.

Assuming the example shown in Fig. 8.7, PMM collects the timestamps and processes them to determine partial and end-to-end latencies.

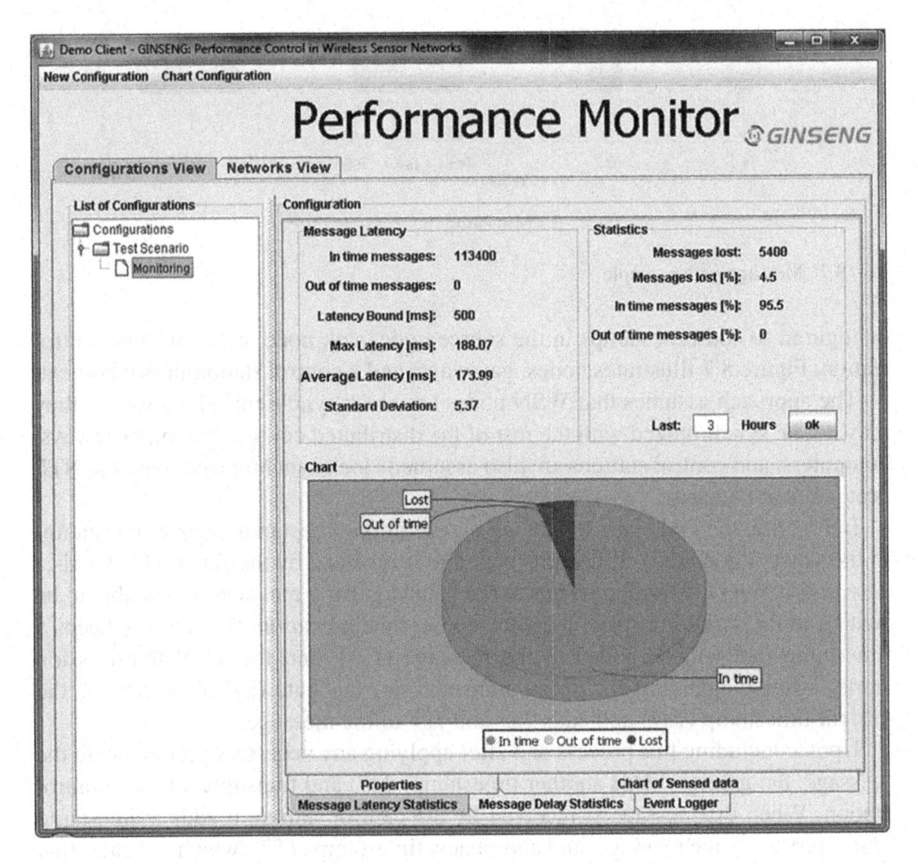

Fig. 8.8 PMM user interface

The following partial latencies are calculated:

- WSN upstream latency (Ts2 − Ts1)
- WSN to gateway interface latency (Ts4 − Ts3)
- Middleware latency (Ts6 − Ts5)
- Control station latency (Ts7 − Ts6)
- End-to-end ((Ts2 − Ts1) + (Ts4 − Ts3) + (Ts6 − Ts5) + (Ts7 − Ts6))

After concluding all computations, PMM stores the following information in the database: source node id, destination node id, type of message, MsgSeqId, [Timestamps], partial latencies, and end-to-end latency. This information is stored for each message, when the second operation mode of debugging is running. When the first operation mode of the debugging component is running, a full report with link-by-link information and end-to-end information is also stored.

The PMM user interface shows operations performance data and alerts users when there is a problem detected by metric exceeds bounds. Statistical information is also shown and is updated for each event that arrives or for each timeout that occurs.

Figure 8.8 shows a screenshot of PMM. We can see how many events (data messages) arrived in-time, out-of-time (with respect to defined bounds), and the

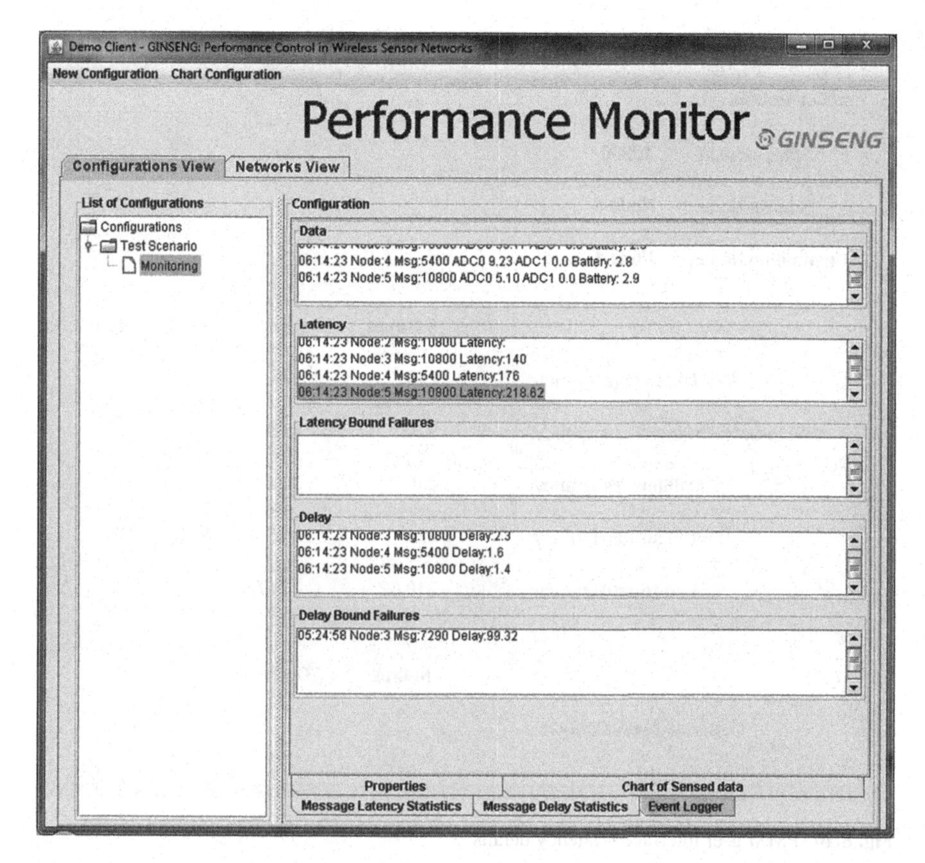

Fig. 8.9 PMM user interface – event logger

corresponding statistical information. This interface also shows a pie chart to give an overall view of the performance.

Figure 8.9 shows the event logger of PMM. This logger shows the details on failed events. A list of failures is shown and the user can select one of each and see all details, including the latency in each part of the distributed control system.

When a problem is reported by the PMM, the user can explore the event properties (e.g., delayed messages, high latencies) and find where the problem occurs. If a problem is found and the debugging report is not available at the PMM, nodes are requested to send their debugging report. If a node is dead, the debugging report is not retrieved and the problem source may be due to the dead node. Otherwise, if all reports are retrieved, the PMM is able to detect the message path and check where it was discarded or where it took longer than expected.

PMM allows users to select one message and see all details, including the latency in each part of the distributed control system. Figure 8.10 shows the details of a message.

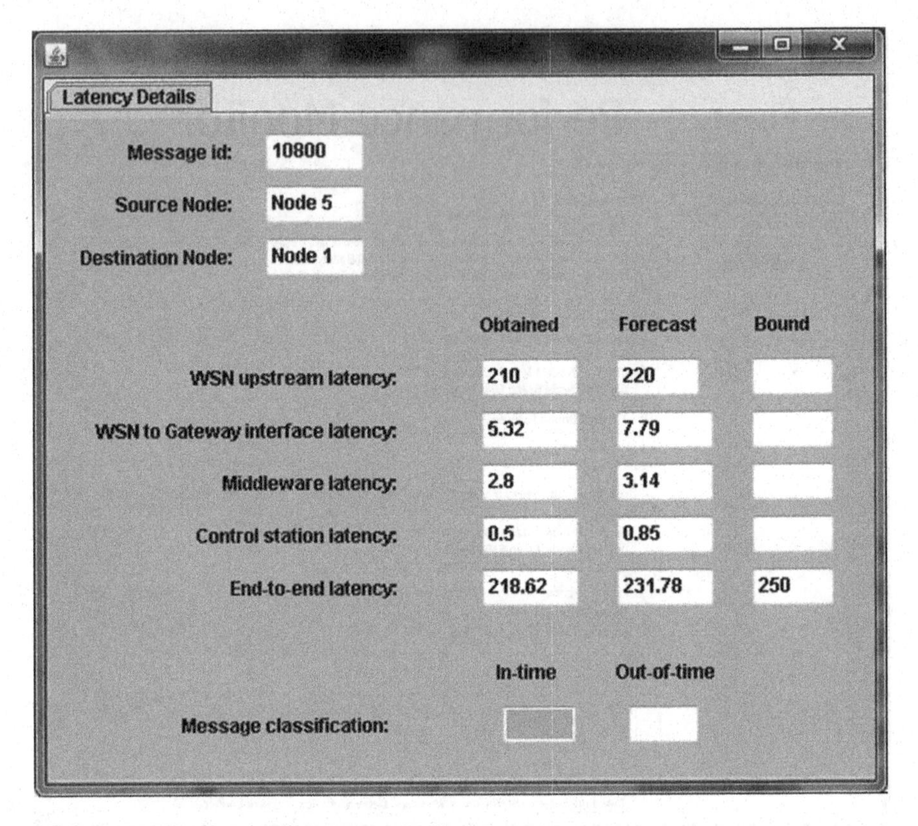

Fig. 8.10 PMM user interface – latency details

In the interface of Fig. 8.10, users can see the latency per parts, as well as the end-to-end latency. This interface also includes information about the forecasted values and bounds. Each bound is specified by the user and can be specified for all parts or only for specific parts. In the example of Fig. 8.10, only an end-to-end bound is defined.

Lastly, this interface also includes information about message classification. This information is filled only when the end-to-end latency bound is defined.

Since DM can be configured to keep information about all messages and operations, reporting those to the PM, PM is able to compute metrics and show report information. Figure 8.11 shows an example of a message traveling from node 1 to node 10. This also allows users to inspect details when they detect a problem by looking at the logger of the Performance Monitor (PM).

In this example, we can see the arriving and departure timestamps, fails, and retransmissions per node. Based on the timestamps collected along the path, PMM computes link latencies for each link, buffering latencies, and the end-to-end latencies. Each link latency is determined through the computation of the difference

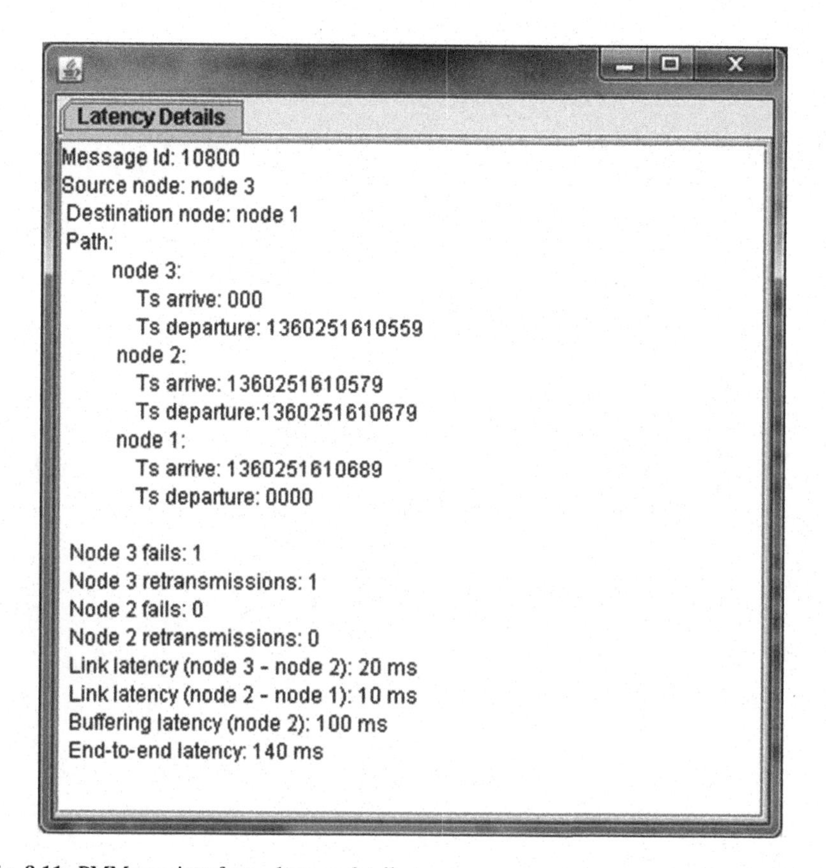

Fig. 8.11 PMM user interface – latency details

between arriving and departure timestamps of receiving node and sending node, respectively. The buffering latency is determined based on the arriving and departure times of each node. This time represents how long the message is kept in a node. Lastly, the end-to-end latency is the sum of all parts of the latencies.

Chapter 9
Application Scenario: An Industrial Case Study

In this chapter, we present, as an application scenario, the results of experimental evaluation of the planning and monitoring approaches in an industrial setup. We will consider a heterogeneous distributed control system where WSN subnetworks coexist with cabled networks.

In the following sections, we will describe an application scenario and a setup that we built to demonstrate the ability of mechanisms and strategies discussed in this book (Sect. 9.1). Based on this application, we will present results comparing observed latencies with forecasts that result from applying planning formulas. Section 9.2 reports results concerning monitoring operation over the setup. Since operation timings may be relative to event occurrence, Sect. 9.3 reports results when considering event occurrence instant. In Sect. 9.4, we show how the algorithm splits the network to meet strict monitoring latency requirements. Section 9.5 shows results concerning the planning algorithm applied to closed-loop operation and Sect. 9.6 shows results concerning the position of downstream slots used to send actuation commands to nodes. In Sect. 9.7 we show results concerning multiple actuators with different timing requirements.

In the planning algorithm described in Chap. 7, to reduce the actuation latency, we add more downstream slots. The number of downstream slots has significant impact in network lifetime when nodes are battery operated. Section 9.8 reports results concerning lifetime versus the number of downstream slots used to meet timing requirements.

Lastly, Sect. 9.9 reports results concerning bounds and debugging tool. We create a simulation environment where we introduce some random delays in the messages, to demonstrate how the debugging tool works and its usability.

For completeness, the charts in this chapter are complemented by Appendix C, where we detail the values in table format.

J. Cecílio and P. Furtado, *Wireless Sensors in Industrial Time-Critical Environments*, Computer Communications and Networks, DOI 10.1007/978-3-319-02889-7_9,

9.1 Application Scenario and Test Bed

There are a lot of applications where distributed control systems with wireless sensors are used. These applications range from military surveillance, in which a large number of sensor nodes are used, to healthcare applications, in which a very limited number of sensor nodes are used. Naturally, these applications have an impact on the specifications of the hardware and software for sensor nodes.

Some researchers have tried to identify possible application scenarios of wireless sensor networks [1–5]. In this section, we will introduce the industrial monitoring and control scenario.

The value of wireless networks is becoming obvious to organizations that need real-time access to information about the environment of their plants, processes, and equipment to prevent disruption [6–8]. Wireless solutions can offer lower system, infrastructure, and operating costs as well as improvement of product quality.

Process Control: In the field of process control, nodes collect and deliver real-time data to the control operator and are able to detect in situ variations in the processes. Nodes may include different sensors and actuators to monitor and control a physical process. They must be able to adjust, for instance, the speed used by a motor, according to the required output. Wireless distributed networks that link different sensors make machine-to-machine communication possible and have the potential to increase the process efficiency in factories. Table 9.1 summarizes the system requirements for environmental application scenarios [6].

Health of equipment monitoring for management and control is one application scenario used in industrial environments. Sensor nodes are continually monitoring to evaluate the "health of machines" as well as their usage. Sensors installed on different machines measure physical properties such as temperature, pressure, humidity, or vibrations. The sensor nodes are able to communicate between each other and send data to the network where the data is processed. When critical values are found, the system immediately sends alarms, making predictive maintenance possible.

Table 9.1 System requirements of process control application scenarios

Requirements	Level
Network lifetime	>8,760 h (1 year)
Scalability	Tens
Time synchronization	Second
Localization	Required (meter)
Security	Low
Addressing	Address-centric
Fault tolerance	Middle
Heterogeneity	Yes
Traffic characteristics	Periodic, queried, event-based
End-to-end delay	1–3 s
Packet loss	Occasional (<5 %)
Traffic diversity	Medium

Table 9.2 System requirements for equipment monitoring application scenario

Requirements	Level
Network lifetime	Forever (access to plant's main power)
Scalability	Planned deployment
Time synchronization	No synchronization
Localization	
Security	Not required
Addressing	Address-centric
Fault tolerance	Middle
Heterogeneity	Yes
Traffic characteristics	Queried, event-based
End-to-end delay	1–3 s
Packet loss	Occasional (<5 %)
Traffic diversity	Low

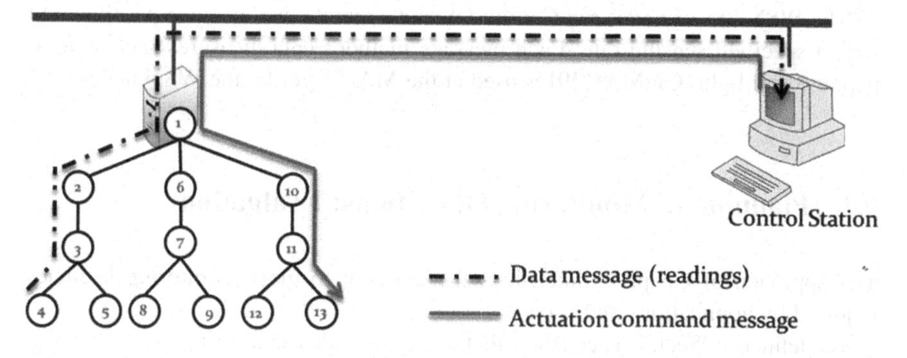

Fig. 9.1 Setup

Table 9.2 summarizes the system requirements for environmental application scenarios [6].

In industrial application contexts, the network is managed by factory employers (no programming expertise) and requires monitoring or closed-loop configurations to control physical processes. Those configurations may change over the time with specific process conditions. Failure of a control loop may cause unscheduled plant shutdown or even severe accidents in process-controlled plants. Users should be able to configure the network operations (monitor and control) easily.

In order to demonstrate capabilities of mechanisms and strategies discussed in this book, we built a test bed and run those mechanisms and strategies. The test bed comprises a control station, a cabled network, one gateway, and a WSN subnetwork. Figure 9.1 shows a sketch of our setup.

The WSN subnetwork includes 12 TelosB nodes organized hierarchically in a 1–1–2 tree and one sink node (composed of one TelosB node and one computer that acts as gateway of the subnetwork). The setup also includes a control station that receives the sensor samples, monitors operations performance using bounds, and collects data for testing the debugging approach.

The control station is a computer with an Intel Pentium D, running at 3.4 GHz. It has 2 GB of RAM and an Ethernet connection. The gateway connecting the WSN subnetwork to the cabled network is another computer with similar characteristics.

All computer nodes (gateway and control station) are connected through Ethernet cables and GigaBit network adapters. The WSN subnetwork is connected to the gateway using the serial interface provided by TelosB nodes. That interface is configured to operate at 460,800 baud/s.

All computers run Linux OS and have specific components developed using java to do specific tasks. For instance, the gateway computer has a gateway software component to read data from the serial interface and send it to the control station and to receive messages destined for the WSN and deliver them through the serial interface. The control station has a software component to perform functionality such as closed-loop control.

The WSN sensor nodes run Contiki OS and generate one message per time unit with a specified sensing rate. Each message includes data measures such as temperature and light. GinMAC [9] is used at the MAC layer by the WSN nodes.

9.2 Planning of Monitoring Operations: Evaluation

The approach of Chap. 7 dimensions the network to meet monitoring latencies required by applications and users.

As defined in Sect. 7.1, consider that a user provides as user inputs:

- The network configuration represented in Fig. 9.1 (in the format exemplified in Appendix A)
- Indication of option 2 of data forwarding rule (each parent collects data from all children and only forwards after receiving from all child nodes)
- Monitoring operation (periodic sensor sampling without event consideration) to run 120 days, at least, over the distributed control system, with a minimum sampling rate of 1 s and a maximum desirable monitoring latency of 200 ms

9.2.1 *Applying TDMA–CSMA Protocols to the Setup*

Based on network configuration, data forwarding rule, latencies, and lifetime requirements, the algorithm creates a schedule for the network. We followed the steps of the algorithm, creating the schedule of Fig. 9.2. This schedule has an epoch with 1 s of length and an inactivity period of 330 ms, determined by Eq. 7.15. It includes sufficient slots for each node to transmit its data upward, where one retransmission slot is added to each transmission slot to enhance reliability. The

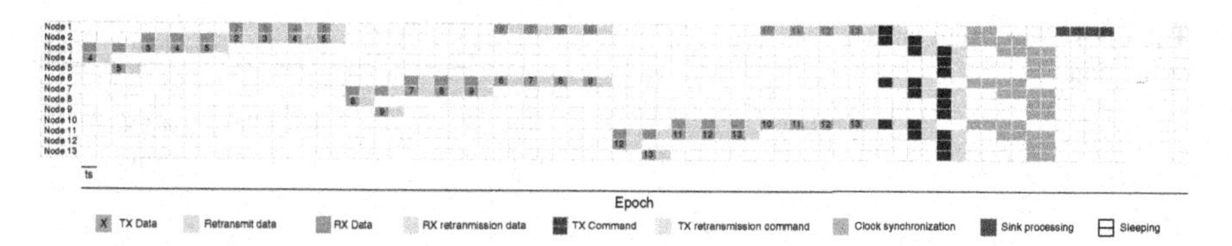

Fig. 9.2 TDMA schedule

Table 9.3 Non-real-time parts characterization [ms]

	t_{Serial}	$t_{\text{Middleware}}$	$t_{\text{Processing}}$
Average	2.64	1.12	0.51
Standard deviation	0.40	0.29	0.12
Maximum	7.79	3.14	0.85
Minimum	1.85	0.67	0.32

schedule also includes transmission slots for sending configuration or actuation commands, slots for time synchronization, and slots for node processing.

In the next subsections, we will verify latency requirements for this resulting schedule using the latency formulas described in Chap. 6, and we will report results from the experimental test bed to compare with the values forecasted.

9.2.2 *Applying Formulas and Testing TDMA–CSMA Setup*

Based on that requirements and applying Eq. 6.14, we obtain

$$\begin{aligned} 200 = &\max\left(t_{\text{WSN}_{\text{AqE}}}\right) + \max\left(t_{\text{WSN}_{\text{UP}}}\right) \\ &+\max\left(t_{\text{Serial}}\right) + \max\left(t_{\text{Middleware}}\right) + \max\left(t_{\text{Processing}}\right) \end{aligned} \tag{9.1}$$

t_{Serial}, $t_{\text{Middleware}}$, and $t_{\text{Processing}}$ were characterized by network testing. The setup ran during 1 h and we collected time statistics. Table 9.3 shows the characterization of t_{Serial}, $t_{\text{Middleware}}$, and $t_{\text{Processing}}$ latencies. All times are given in milliseconds.

Replacing those latencies in Eq. 6.14 and considering $\max(t_{\text{Aq}})$ equal to 20 ms (it was determined by node testing), we need to determine the amount of latency due to WSN subnetwork ($\max(t_{\text{WSN}_{\text{UP}}})$). It is obtained by analysis of the data flow and the schedule applied to the WSN subnetwork.

Since the setup is organized as a tree hierarchy with three levels, $\max(t_{WSN_{UP}})$ will assume three different values. Depending on the node position and according to the schedule of Fig. 9.2, $\max(t_{WSN_{UP}})$ can be 20, 100, or 160 ms.

Table 9.4 Maximum latency estimation

Level	Latency [ms]
1	51.78
2	131.78
3	191.78

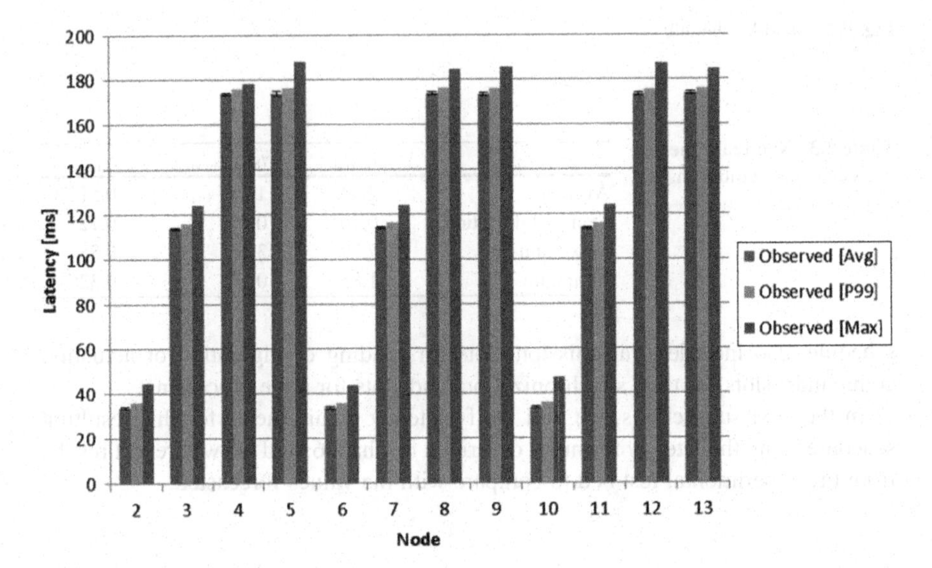

Fig. 9.3 Monitor latency per node

Therefore, applying Eq. 6.14 with the values of Table 9.3, the monitoring latency for nodes at level 1 can be predicted as

$$\max\left(\text{Monitoring}_{\text{Latency}}\right) = 40 + 7.79 + 3.14 + 0.85 = 51.78\,\text{ms} \quad (9.2)$$

The same equation is applied to the other levels, resulting in the times shown in Table 9.4. All nodes at the same level have the same forecast.

To assess the latency model, we tested the network layout resulting from the algorithm (Fig. 9.2) and compared the latency results with the expected latencies calculated in the previous subsection, which were given by applying the latency formulas of Chap. 6. Figure 9.3 shows statistical information of latency per node, gathered from an experiment that ran for 3 days.

From Fig. 9.3, we can see that nodes at the same level have similar latencies. Figure 9.4 shows statistical information of latency (per level), as well as the values corresponding to the prediction given by the planning formulas (Table 9.4).

From Fig. 9.4, we can conclude that the planning approach predicts well for this setup. The observed maximum value gathered during the test is always below the prediction. It is near, but below the planned maximum.

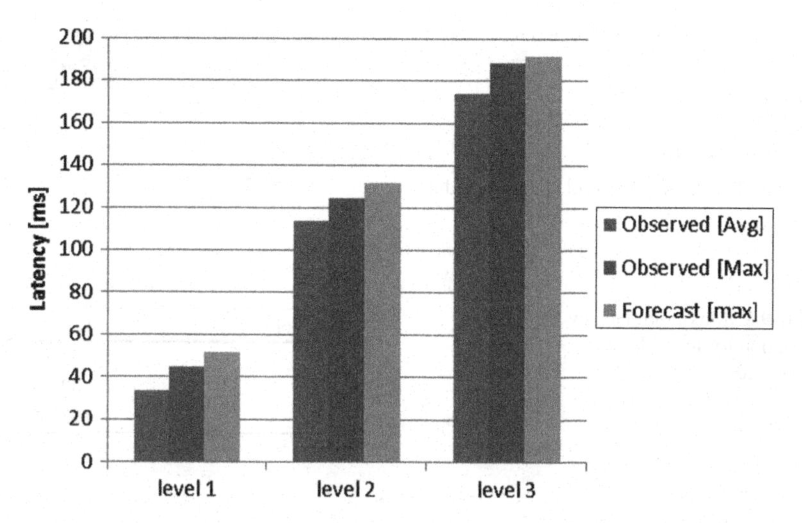

Fig. 9.4 Monitor latency per level with forecast

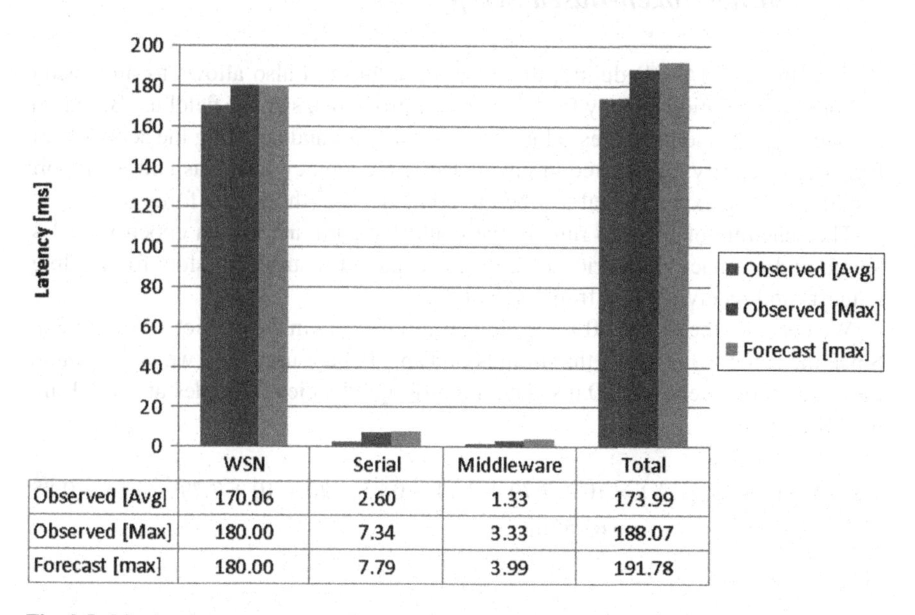

	WSN	Serial	Middleware	Total
Observed [Avg]	170.06	2.60	1.33	173.99
Observed [Max]	180.00	7.34	3.33	188.07
Forecast [max]	180.00	7.79	3.99	191.78

Fig. 9.5 Monitor latency per network part

The latencies shown in Fig. 9.4 can be decomposed into the following latencies: WSN latency, serial interface latency, and the middleware latency. Figure 9.5 shows those latencies for a node in the third level.

The WSN part is, in average, 170.06 ms, but it can grow up to 180 ms. This maximum value agrees with the prediction (forecasted values). Concerning other parts such as serial and middleware latencies, the obtained maximum values show a little difference to the predicted values, but all of them are below the prediction.

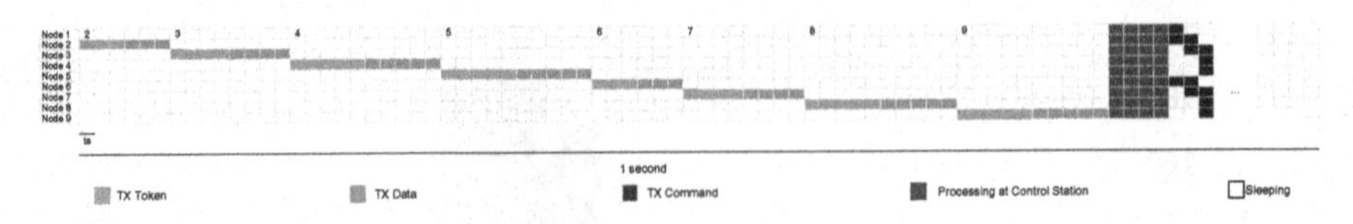

Fig. 9.6 Token-based control station schedule

Table 9.5 Maximum latency estimation for token-based end-to-end network

Level	Latency [ms]
1	63.56
2	83.56
3	103.56

9.2.3 Applying Formulas and Testing for the Token-Based Setup

Besides the TDMA scheduling, the planning algorithm also allows to dimension and predict operation latency for token-based protocols, such as fieldbus. Based on the same requirements expressed in the previous case and applying the schedule of Fig. 9.6, we verify latency requirements using the latency formulas and we report results from the experimental test bed to compare with the values forecasted.

The schedule of Fig. 9.6 runs in the control station and has an epoch with 1-s length and an inactivity period of 292 ms. It includes sufficient slots for sending requests and receiving data from each node.

We consider the same 200-ms timing requirement as in the previous experiment. Since the setup is the same, the timings of Table 9.3 are used, and we again bound the acquisition latency as 20 ms. The monitoring latencies for nodes at level 1 are then (Eq. 6.16)

$$\begin{aligned}\max\left(\text{Monitoring}_{\text{Latency}}\right) &= 10+7.79+3.14+0.85+20+10+7.79+3.14+0.85\\ &= 63.56\,\text{ms}\end{aligned} \tag{9.3}$$

The same equation is applied to the other levels, resulting in the times shown in Table 9.5. All nodes at the same level have the same forecast.

To assess these latency predictions, we tested the network with the schedule of Fig. 9.6 and compared the latency results with the latencies predictions. Figure 9.7 shows statistical information of latency per node, gathered from an experiment that ran for 1 h.

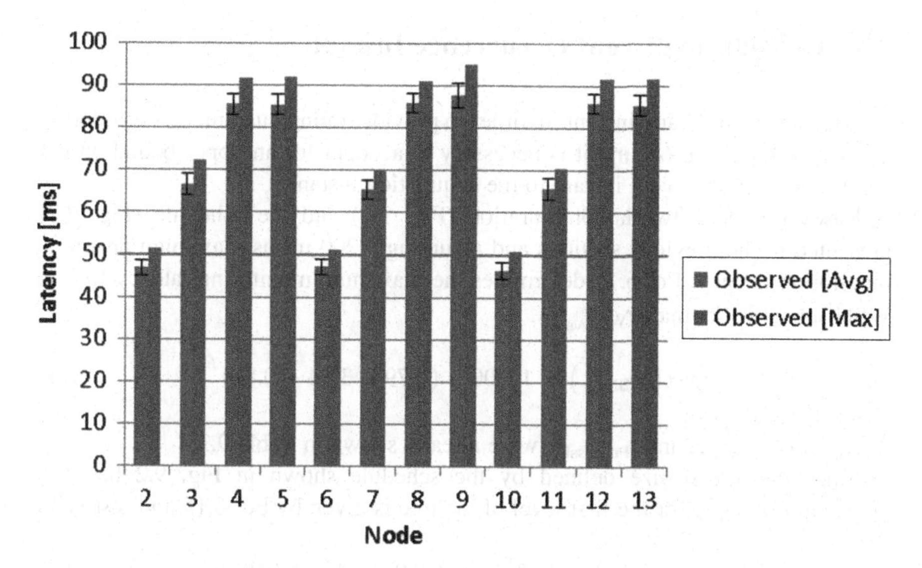

Fig. 9.7 Monitor latency per node

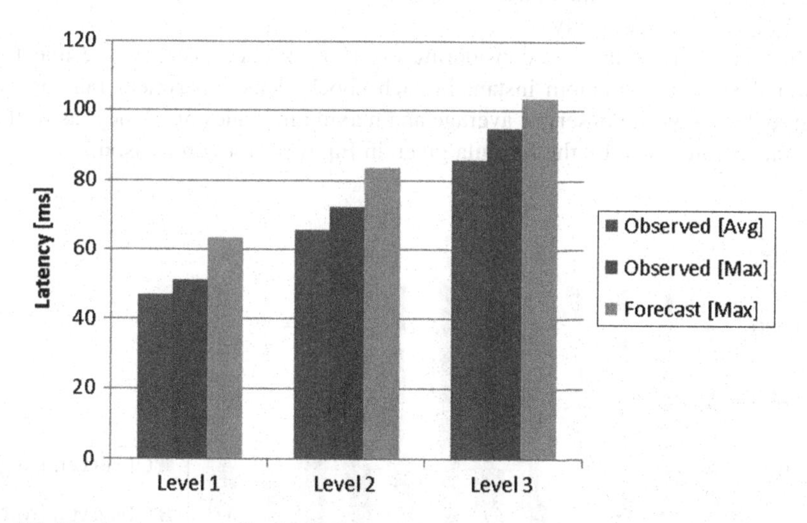

Fig. 9.8 Monitor latency per level with forecast

From Fig. 9.7, we can see that nodes at the same level have similar latencies. Figure 9.8 shows statistical information of latency (per level), as well as the values corresponding to the prediction given by the latency model formulas (Table 9.5).

From Fig. 9.8, similar to the previous case, we can conclude that the latency model predicts well for this setup. The observed maximum value gathered during the test is always below the prediction.

9.3 Considering Event Occurrence Instant

Events may occur in any instant. In order to provide timing guarantees, considering the instant the event occurs, it is necessary to account for an upper bound on the extra time from the event instant to the acquisition instant.

Based on the network configuration (Fig. 9.1) and the schedule (Fig. 9.2) presented in the previous sections and assuming 1,200 ms as maximum desirable monitoring latency, Eq. 6.15 determines the maximum monitoring latency for the WSN subnetwork $\left(\max\left(t_{\mathrm{WSN}_{\mathrm{AqE}}}\right)\right)$.

$$\max\left(t_{\mathrm{WSN}_{\mathrm{AqE}}}\right) = 1,200 - (7.79 - 3.14 - 0.85) \tag{9.4}$$

t_{Serial}, $t_{\mathrm{Middleware}}$, and $t_{\mathrm{Processing}}$ were already shown in Table 9.3.

Since the epoch size defined by the schedule shown in Fig. 9.2 has 1 s (1,000 ms), $t_{\mathrm{WSN}_{\mathrm{AqE}}}$ for the first level of the tree is given by Eq. 6.15 and results in

$$\max\left(t_{\mathrm{WSN}_{\mathrm{AqE}}}\right) = 1,000 + 20 + 0 + 20 = 1,040\,[\mathrm{ms}] \tag{9.5}$$

Applying the same equation to the other levels, we obtain 1,120 and 1,180 ms for levels 2 and 3, respectively.

In order to test latencies considering event occurrence instance, we inject an external event at a random instant in each epoch. This experiment ran for 2 h. Figure 9.9 shows the observed average and maximum values of latency, as well as the values forecasted by the formula given in Eq. 6.14, for comparison.

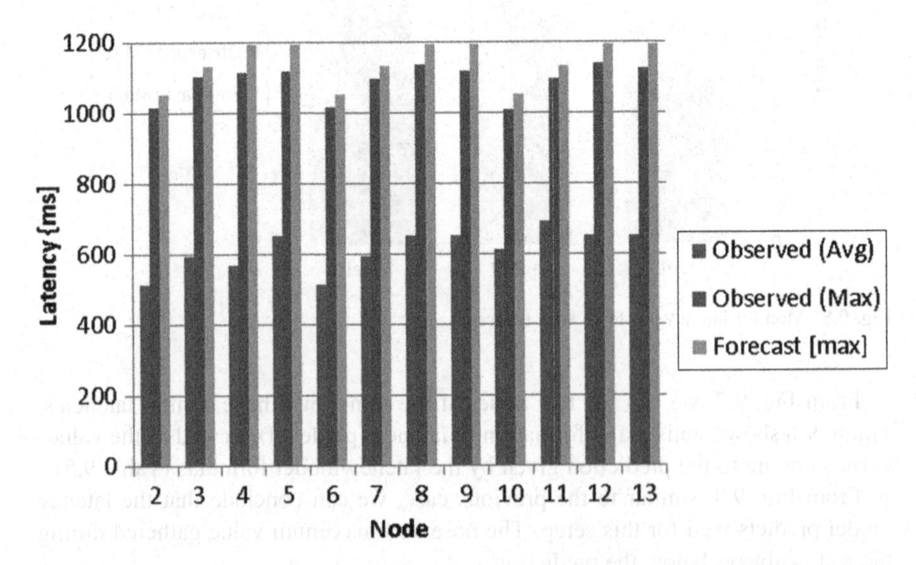

Fig. 9.9 Event detection latency per node (observed and forecasted)

From Fig. 9.9 we can see that an event takes, in average, half of the epoch to be identified in the control station. The observed maximum values are similar to the forecasted maximum latencies, as expected.

9.4 Planning with Network Splitting

The approach proposed dimensions the WSN network to meet latency requirements. For instance, considering the same setup of the previous experiment, if a user specifies 500 ms as maximum latency, event occurrence should be reported within that timing constraint. Since the forecast of maximum latency was much larger (1,200 ms), the network should be resized to define an epoch which meets the latency requirement.

Applying the planning algorithm, the initial network is divided into three subnetworks where each includes one branch only, resulting in the schedule shown in Fig. 9.10.

To evaluate the resulting subnetwork and schedule, we built a test bed and ran it during 2 h. Figure 9.11 shows the statistical and forecasted values of latency.

From Fig. 9.11 we can see that the new layout guarantees the timing constraints and the forecast values agree with our test bed (maximum latency bounded by 500 ms).

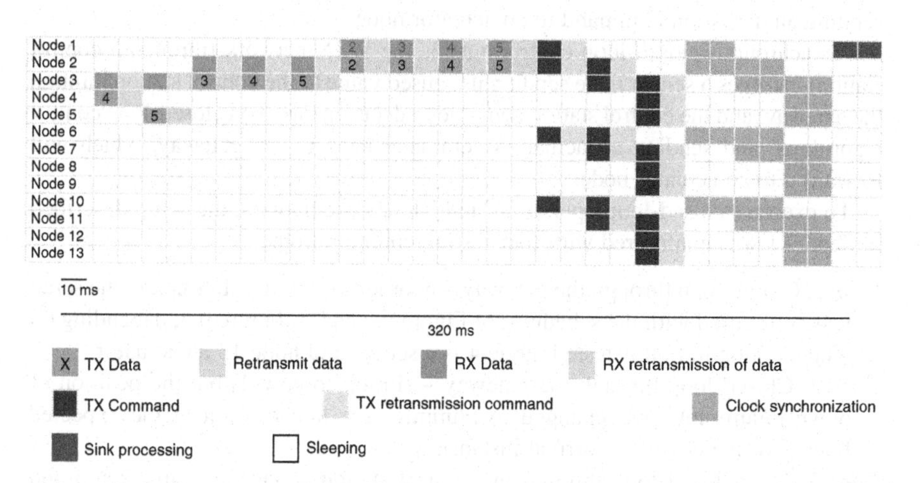

Fig. 9.10 TDMA schedule to meet the event latency

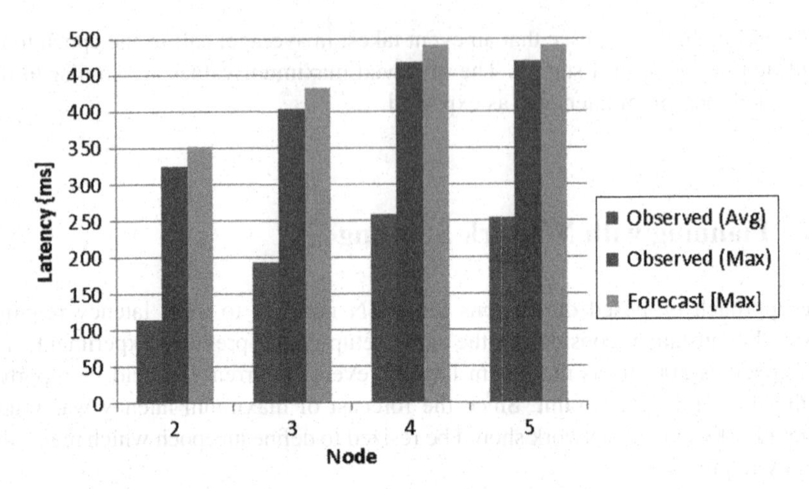

Fig. 9.11 Event detection latency (observed and forecasted)

9.5 Planning of Closed-Loop Operation: Evaluation

Closed-loop operations can be configured to be processed inside the WSN subnetwork (through the gateway) or outside the WSN subnetwork (through a control station). Asynchronous closed loops through the gateway involve sensor nodes sending sensed values to the gateway, the gateway evaluating a threshold and sending an actuation command to an actuator node.

Asynchronous closed loop control outside the WSN network (through a control station) involves a sensor node sending its sensed value to the control station through the gateway, and the control station computes a decision based on closed-loop control algorithms, and sending an actuation command back to the gateway, which will forward it to an actuator node.

To exercise closed-loop operations and latency predictions, the setup described in Sect. 9.1 was configured with four testing configurations:

Case 1: Closed loop through the gateway – in order to test this alternative, the setup was configured with the schedules of Figure 9.2 and/or Figure 9.6, depending on which configuration is tested; node 4 is a sensor and node 13 an actuator.

Case 2: Closed loop through the gateway – similar to case 1, but the position of downstream slots was changed to minimize the latency (put at the expected bound for the command arrival instant).

Case 3 & 4: Closed loop through the control station – with the same setup and schedules used in cases 1 and 2, we change the position of the supervision control logic to run at the control station and analyze the result latencies.

Table 9.6 shows the values for the four setups and asynchronous closed-loop control. The values resulted from applying the latency model defined in Chap. 6, marked as

Table 9.6 Closed-loop latencies [ms]

Configuration			WSN up	Serial	Middleware	Processing	Wait for TX slot	WSN down	CMD processing	Total
Case 1	Wireless TDMA, wired CDMA	Test bed result	180	7.3	–	0.86	380	60	0.61	628.8
		Forecast	180	14.2	–	1	380	60	1	636.2
	Wireless and wired token-based end-to-end	Test bed result	30.6	6.8	–	0.57	498.3	27.3	1.6	565.2
		Forecast	60	14.2	–	1	510	30	2	617.4
	Wireless TDMA and wired token-based	Test bed result	180	7.3	–	0.86	380	60	0.61	628.8
		Forecast	180	14.2	–	1	380	60	1	636.2
Case 2	Wireless TDMA, wired CDMA	Test bed result	180	6.2	–	0.83	13	60	0.62	260.7
		Forecast	180	14.2	–	1	20	60	1	276.2
	Wireless and wired token-based end-to-end	Test bed result	30.2	7.5	–	0.67	10	26.8	1.78	77
		Forecast	60	14.2	–	1	10	30	2	117.2
	Wireless TDMA and wired token-based	Test bed result	180	6.2	–	0.83	13	60	0.62	260.7
		Forecast	180	14.2	–	1	20	60	1	276.2
Case 3	Wireless TDMA, wired CDMA	Test bed result	180	7.3	5.85	0.96	380	60	0.91	635
		Forecast	180	14.2	10.1	1	380	60	1	646.3
	Wireless and wired token-based end-to-end	Test bed result	33.6	6.6	5.4	0.87	493.3	27.3	1.6	568.7
		Forecast	60	14.2	10.1	1	510	30	2	627.3
	Wireless TDMA and wired token-based	Test bed result	180	7.3	28.85	0.96	380	60	0.91	658
		Forecast	180	14.2	40.1	1	380	60	1	676.3
Case 4	Wireless TDMA, wired CDMA	Test bed result	180	7.1	4.85	0.86	11	60	0.71	264.5
		Forecast	180	14.1	10.1	1	20	60	1	286.2
	Wireless and wired token-based end-to-end	Test bed result	31.9	8.1	5.2	0.7	9.3	25.9	1.68	82.8
		Forecast	60	14.2	10.1	1	10	30	2	127.3
	Wireless TDMA and wired token-based	Test bed result	180	7.1	34.82	0.86	11	60	0.71	294.5
		Forecast	180	14.1	40.1	1	20	60	1	316.2

"Forecast," while the remaining values represent the average values obtained from experimental validation.

Each time measure of Table 9.6 represents the total amount of time consumed per part. For instance, considering a closed-loop operation with decision logic in the gateway, a data sample must travel twice the serial interface. So, the latency shown in Table 9.6 represents the sum of travels.

In a similar way, the middleware latency for wireless TDMA and wired token-based network is the sum of t_{Gateway}, t_{LAN}, and $t_{\text{Wait Token Slot}}$.

From Table 9.6, we conclude that the observed latencies were always within the bounds defined by the forecasted maximum latencies, which allows concluding that the latency model defined in Chap. 6 predicts well operation latencies.

9.6 Adding Downstream Slots Equally Spaced in the Epoch

As discussed in Chap. 6, command latencies can be decreased by adding equally spaced downstream slots. To guarantee latency reduction, the number of downstream slots can be increased.

Figure 9.12 shows the influence of the number of downstream slots in the end-to-end closed-loop latency. In this experiment the network shown in Fig. 9.1 and the schedule of Fig. 9.2 were used. The closed-loop decision was configured to run in the control station and the network was configured to send one sample per second. Six 40-min experiments were ran corresponding to different number of downstream slots (1, 2, or 4) and the two closed-loop alternatives (asynchronous and synchronous).

Figure 9.12a shows results concerning asynchronous closed-loop control, while Fig. 9.12b corresponds to synchronous closed-loop control.

From Fig. 9.12a we can see that asynchronous closed loops are able to catch the downstream slot in the same epoch where the sensing happened. This is seen by comparing the observed [max] with "forecast [max] *if catches downstream slot.*" From Fig. 9.12b, we can conclude that if one downstream slot is used per epoch, the command waits a maximum of one epoch (1,000 ms) to be transmitted to the target node. In average, a command is delivered in 1/2 of the epoch time plus travel time. In Fig. 9.12a, b, if more slots are used, $t_{\text{Wait for TX slot}}$ is successively reduced to half for each added slot.

9.7 Multiple Closed Loops

In this section we consider the case of multiple closed loops with decision logic in the control station, with different timing requirements. In this case, the number of equally spaced downstream slots should be dimensioned to meet the most restrictive timing requirement. For instance, consider the setup of Fig. 9.1 and that we

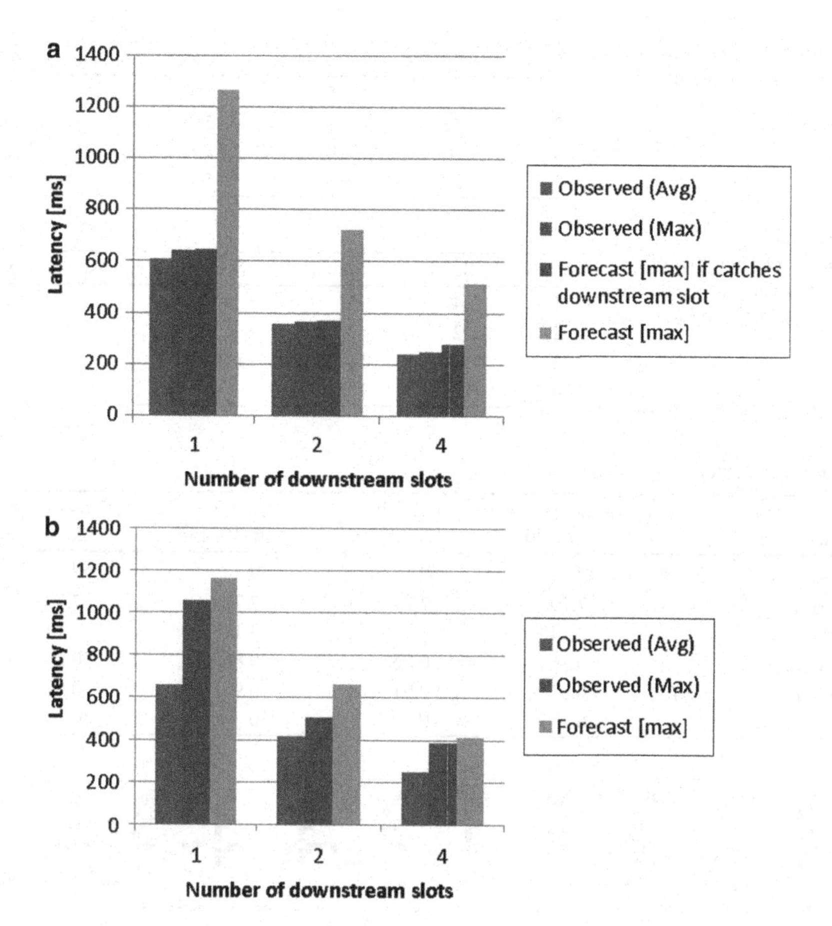

Fig. 9.12 Closed-loop latency versus number of downstream slots (**a**) asynchronous, (**b**) synchronous

have four actuators placed in nodes 5, 7, 10, and 13, which must be controlled according to the sensor data collected by sensors 4, 2, 6, and 12, respectively, with the time requirements shown in Table 9.7.

From Table 9.7, we can see that the strictest closed-loop time is 150 ms. Applying Eq. 7.10, we obtain the number of downstream slots (Table 9.8) needed to meet each latency of Table 9.7. Table 9.9 shows the values of each parameter used in Eq. 7.10 to determine how many slots are needed for each case.

From Table 9.8 we conclude that 10 downstream slots are needed to meet the closed-loop time constraint for all configurations of Table 9.7. Figure 9.13 shows the obtained schedule that meets all closed-loop times. This schedule results from the planning algorithm, when we configure it to place the downstream slots equally spaced in the epoch.

Figure 9.14 shows the closed-loop latencies that were observed, the forecasted values and the maximum admissive latency indicated by the user for the closed loops (user requirement).

Table 9.7 Closed-loop latency requirements

Case	Sensor	Actuator	Closed-loop latency [ms]
1	4	5	500
2	2	7	250
3	6	10	150
4	12	13	500

Table 9.8 Number of downstream slots required to meet latency requirements

Case	Sensor	Actuator	Number of downstream slots
1	4	5	5
2	2	7	7
3	6	10	10
4	12	13	5

Table 9.9 Latency parameters [ms]

Parameter	Case 1: [4, 5, 500]	Case 2: [2, 7, 250]	Case 3: [6, 10, 175]	Case 4: [12, 13, 500]
$\max(t_{\mathrm{WSN}_{\mathrm{AqE}}})$	180	40	40	180
$\max(t_{\mathrm{Serial}})$	7.79	7.79	7.79	7.79
$\max(t_{\mathrm{Middleware}})$	3.14	3.14	3.14	3.14
$\max(t_{\mathrm{Processing}})$	0.86	0.86	0.86	0.86
$\max(t_{\mathrm{CMD\ Processing}})$	0.91	0.91	0.91	0.91
$\max(t_{\mathrm{WSN}_{\mathrm{Down}}})$	30	20	10	30

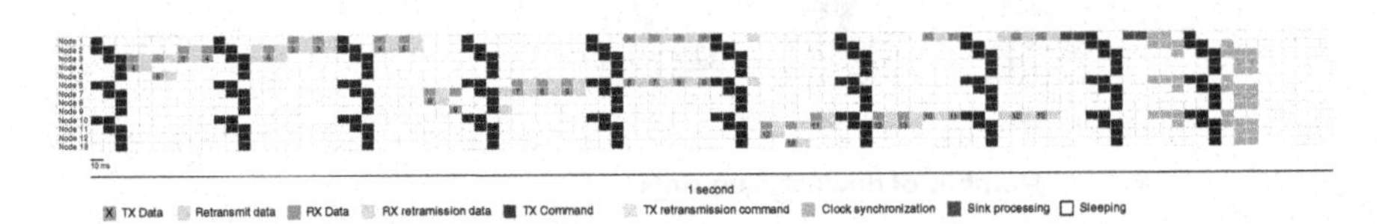

Fig. 9.13 TDMA schedule that meets the strictest closed-loop latency

Figure 9.14 shows that all forecast [max] bounds were met in the experiment. The total latencies vary from case 1 to case 4 because the sensors and actuators are in different positions of the network layout. In cases such as case 1 and case 4, more than one downstream slot were passed by before the actuation command was determined and ready to go down.

From Fig. 9.14, we can also conclude that required timings were always met if actuation commands catch the first downstream slot ahead after receiving the sensed data by the sink node. Actuator 10 has the strictest latency requirement (150 ms) but, as we can see in the figure, that restriction is met. However, the figure also shows that if actuation commands did not catch the first downstream slot ahead and the second downstream slot would be caught instead, the latency requirement for case 3 would not be met. The command would take about 180 ms to be delivered to the actuator at node 10, while the requirement was 150 ms.

Figure 9.15 shows the results concerning latency for synchronous closed-loop control.

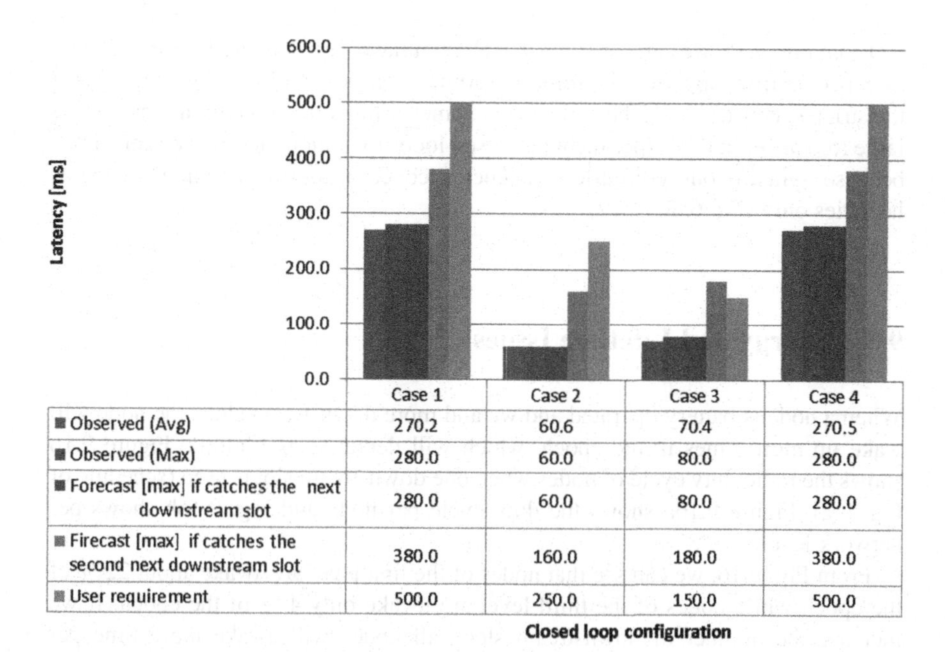

	Case 1	Case 2	Case 3	Case 4
Observed (Avg)	270.2	60.6	70.4	270.5
Observed (Max)	280.0	60.0	80.0	280.0
Forecast [max] if catches the next downstream slot	280.0	60.0	80.0	280.0
Firecast [max] if catches the second next downstream slot	380.0	160.0	180.0	380.0
User requirement	500.0	250.0	150.0	500.0

Fig. 9.14 Asynchronous closed-loop latency for all configurations

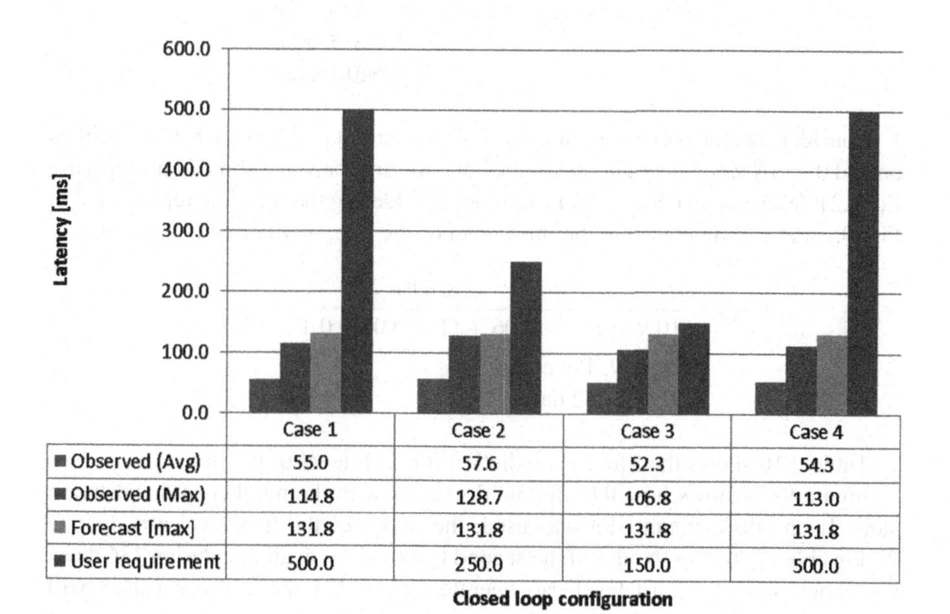

	Case 1	Case 2	Case 3	Case 4
Observed (Avg)	55.0	57.6	52.3	54.3
Observed (Max)	114.8	128.7	106.8	113.0
Forecast [max]	131.8	131.8	131.8	131.8
User requirement	500.0	250.0	150.0	500.0

Fig. 9.15 Synchronous closed-loop latency for all configurations

From Fig. 9.15, we can conclude that all commands are delivered within 132 ms, as expected from applying the forecast formulas of the algorithm. This means that the strictest constraint (150 ms of case 3) is met, while other configurations show a large tolerance. In this experiment all closed loops took more or less the same times because synchronous closed-loop latencies concern control station to actuator latencies only (Eq. 6.8).

9.8 Energy and Lifetime Issues

When a node is battery operated and we add more downstream slots, the node will wake up more times in the epoch, which will decrease its lifetime. Figure 9.16 shows the radio duty cycle of nodes when one downstream slot is used (schedule of Fig. 9.2). Figure 9.16a shows the duty cycle per node and Fig. 9.16b shows per network level.

From Fig. 9.16, we can see that nodes of the first level are awake about 22 % of the epoch while nodes of the third level are awake only 4 % of the epoch. If we increase the number of downstream slots, all nodes will awake more time per epoch, which will decrease the lifetime.

Assuming that a node has two batteries, each with 1.5 V and 800 mAh, the total available energy can be calculated as

$$\begin{aligned}\text{Battery Energy [Joules]} &= \text{V} \cdot \text{Ah} \cdot 3{,}600 \\ &= 3{*}0.8{*}3{,}600 \\ &= 8{,}640 \text{ Joules}\end{aligned}$$

Considering that a node consumes a constant current of 0.9 mA when the radio is on and 0.1 mA when the radio is off, the node lifetime can be estimated by applying Eq. 7.21 described in Chap. 7. For example, considering the schedule represented in Fig. 9.2 and a node placed at the third level of the tree, its lifetime is given by

$$\begin{aligned}T_{\text{Total}} &= \frac{8{,}640}{\left(0.9 \times 10^{-3} \cdot 0.06 + (1 - 0.06) \cdot 0.1 \times 10^{-3}\right) \cdot 3} \\ &= 19{,}459{,}459 \text{ cycles} \\ &= 2{,}252{,}252 \text{ days}\end{aligned} \tag{9.6}$$

Table 9.10 shows the lifetime prediction for each level of the tree.

Figure 9.17 shows how the lifetime decreases with the number of downstream slots. If one downstream slot was used, the nodes of the first level can operate during about 120 days. Nodes of the second level will be available during 136 days, while nodes of the third level can operate during 225 days. These values will

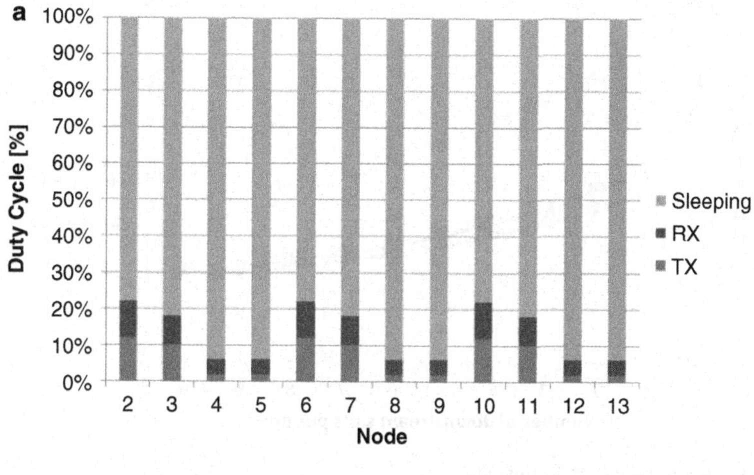

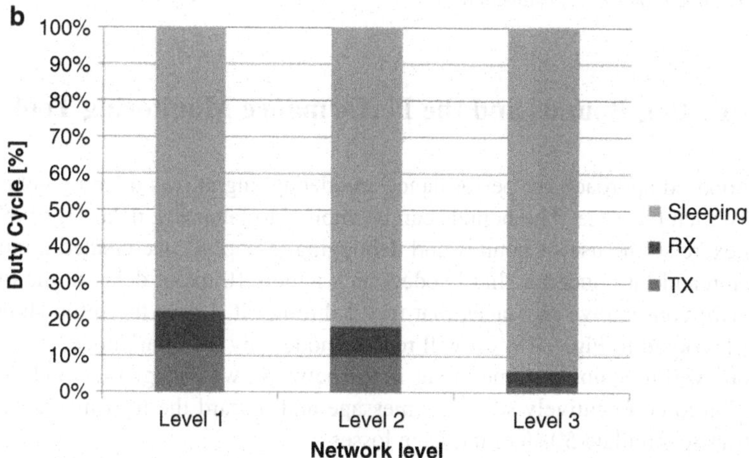

Fig. 9.16 Radio duty cycle (**a**) duty cycle per node, (**b**) duty cycle per network level

Table 9.10 Lifetime prediction

Level	Node lifetime [days]
1	1,207,729
2	1,366,120
3	2,252,252

decrease according to the number of downstream slots. For instance, if we add 10 downstream slots, we will reduce the configuration or actuation command latency, but the lifetime of the node will decrease drastically. In this situation, a node of the first level will only be available for 59 days.

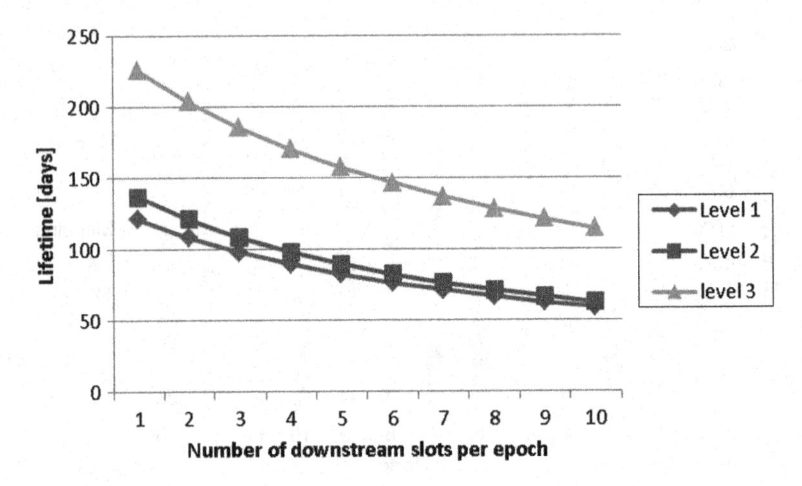

Fig. 9.17 Radio duty cycle estimation

9.9 Testing Bounds and the Performance Monitoring Tool

The proposed approach for performance and debugging allows defining bounds to classify each message. The bounds can be applied to latencies, delays, or both.

To exercise the use of bounds and debugging (Chap. 8), we created a monitor operation and introduced a "liar" node which injects 10 ms of delay in the first of every two consecutive messages that travel through it. Using the setup shown in Fig. 9.1, copied to Fig. 9.18, we will replace node 3 by the "liar" node.

Moreover, to simulate some losses in the network, we changed the node 4 configuration to consecutively send one message and discard the next message. This allows us to simulate 50 % of message losses.

Figures 9.19 and 9.20 show the results concerning message delays. Figure 9.19 reports values concerning delay without the "liar" node, while Fig. 9.20 reports delays after replacement of node 3 by the "liar" node.

From Fig. 9.19, we can conclude that consecutive messages arrive at the control station, in average, within 0.5–2 ms. This value can grow up to a maximum of 8 ms.

After introducing our "liar" node and running the monitoring operation for 24 h, we obtained the chart of Fig. 9.20. This figure shows that the delay of nodes 4 and 5 increased. These nodes send their messages to the control station passing through the "liar" node, which is node 3. In this case, we can see that the delay of two consecutive messages increased, in average, to 12 ms, and up to a maximum of 20 ms.

Using the PM described in Chap. 8, we can also define bounds for the message delay. Assuming that each message should arrive at the control station within a maximum delay of 10 ms, we can define a delay bound and analyze the results.

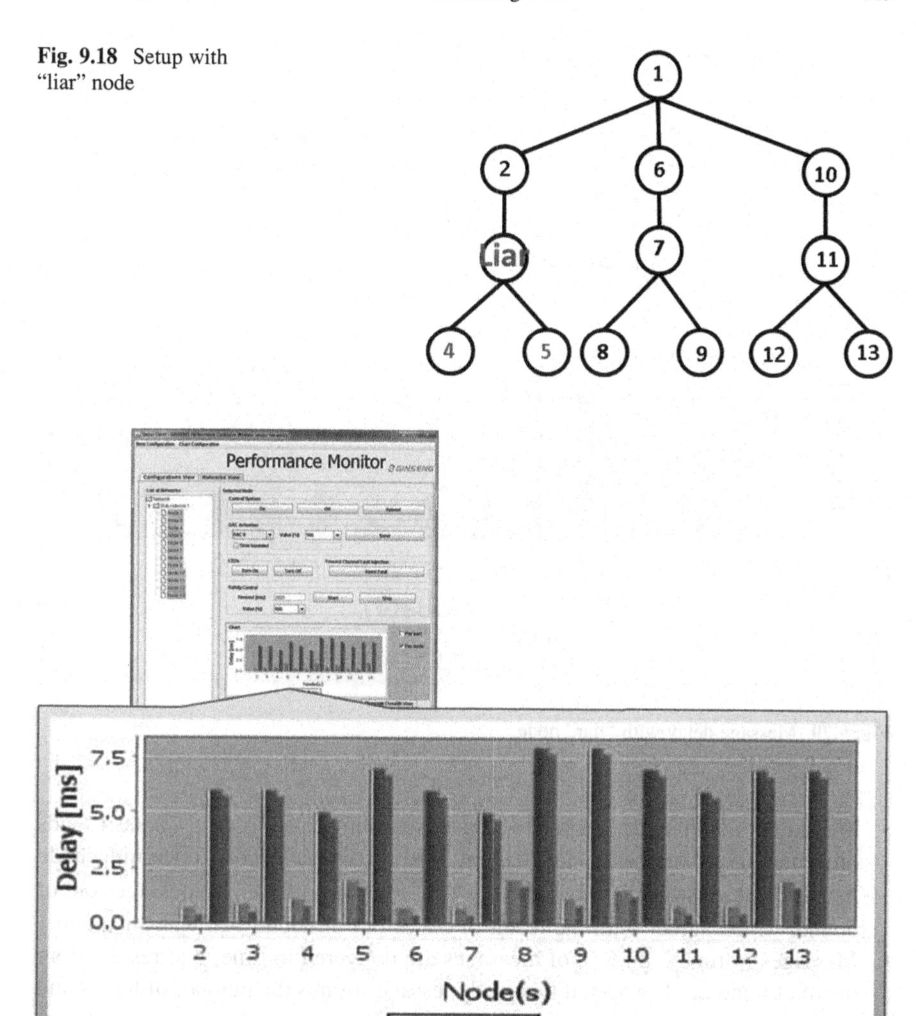

Fig. 9.18 Setup with "liar" node

Fig. 9.19 Message delay without "liar" node

Figure 9.21 shows the percentage of messages classified into each category (in-time, out-of-time, lost), according to a delay bound of 10 ms and lost timeout of 1 s (the timeout when a message is considered lost).

From Fig. 9.21, we conclude that 88.6 % of the messages are delivered within the bound, but 6.8 % are delivered out of bounds and 4.5 % are lost. These numbers are as expected:

- Messages lost – there are 12 nodes sending data messages, node 4 fails one in every two messages. That results in $\frac{1}{12^{*}2}$ losses, which agrees with the result of 4.5 % losses that was obtained.

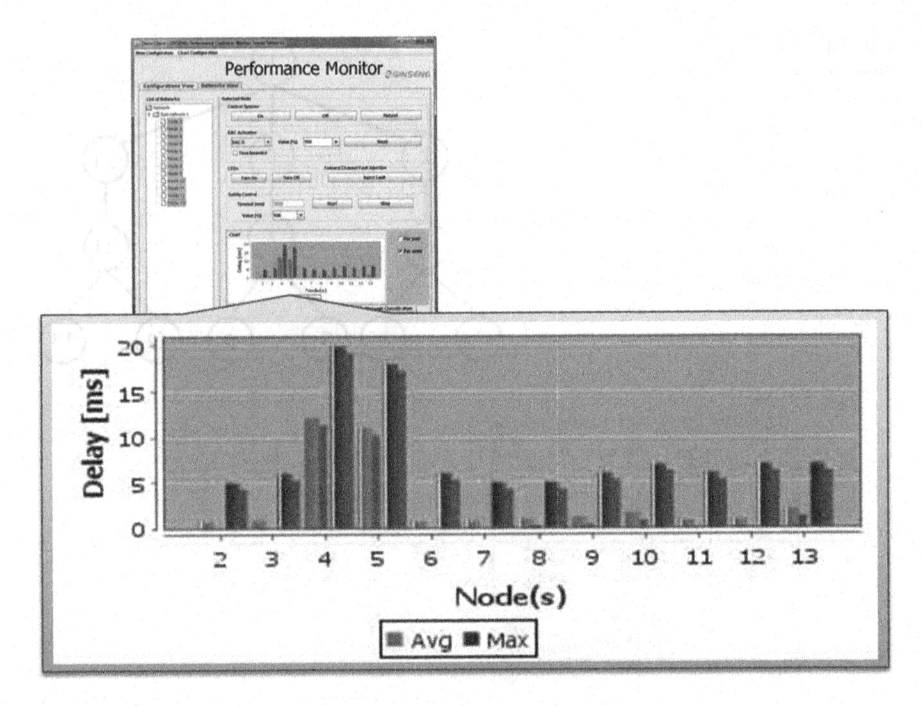

Fig. 9.20 Message delay with "liar" node

- Messages out of bound – there are 12 nodes sending data messages, node 4 sends only half of its messages and half of them arrive delayed. Concerning node 5, half of its messages arrive delayed. That results in $\frac{1}{12^*3} + \frac{1}{12^*2}$ messages out of bound, which agrees with the result of 6.8 % out of bound that was obtained.
- Messages in time – 88.6 % of messages are delivered in time, that results from the total number of expected received messages minus the number of losses and out of bound $\left(1 - \left(\frac{1}{12^*2} + \frac{1}{12^*3} + \frac{1}{12^*2}\right)\right)$.

As described in Chap. 8, the user can also explore event properties (in this case, delays) and find where the problem occurred. For instance, the user interface includes per node evaluation such as the one in Fig. 9.22, which shows which node(s) is failing.

From Fig. 9.22 we can conclude that node 4 is responsible for the losses represented in Fig. 9.21. It is losing 50 % of the messages, as expected.

Figure 9.22 also shows that the delay bound is not met by nodes 4 and 5. In this case, further debugging allows the user to identify the path of each message and to check where it took longer than expected.

For instance, if we explore the path and delay parts of node 5, we can conclude that messages sent by that node are waiting, in average, 10 ms in the transmission queue of node 3 (our "liar" node), which is greater than the expected average delay value of 2 ms for node 3 sending messages to the control station seen in Fig. 9.19.

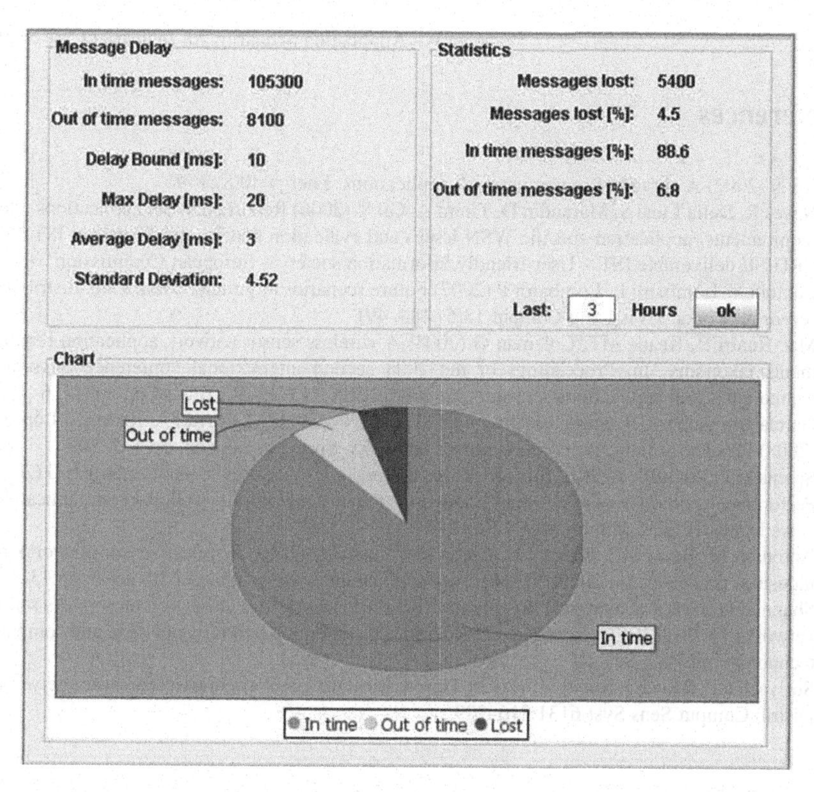

Fig. 9.21 Message classification according to delay bound

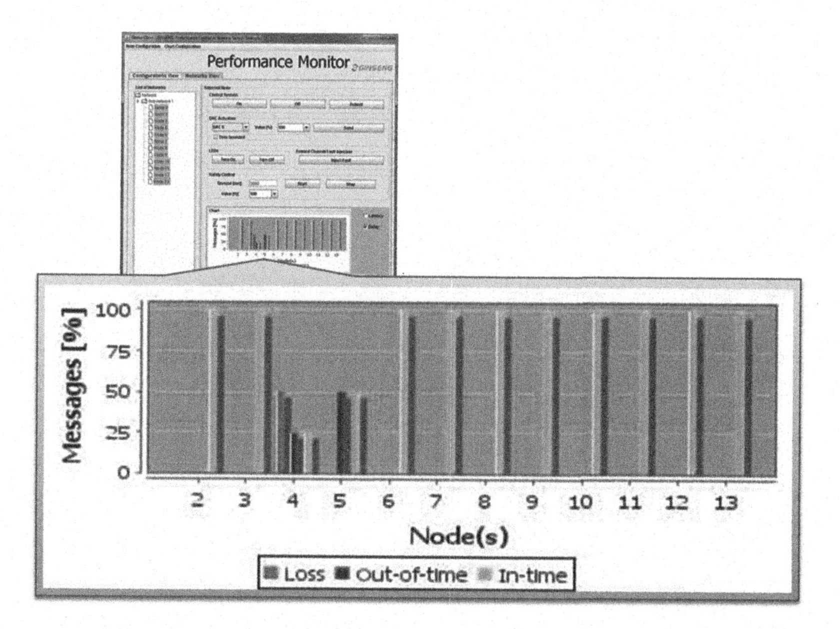

Fig. 9.22 Message classification according to delay bounds per node

References

1. Xu N (2002) A survey of sensor network applications. Energy 40(8):1–9
2. Neves R, Della Luna S, Marandin D, Timm A, Gil V (2006) Report on WSN applications, their requirements, application-specific WSN issues and evaluation metrics. In: European IST NoE CRUISE deliverable IST – User-friendly information society – European Commission
3. Cantoni V, Lombardi L, Lombardi P (2007) Future scenarios of parallel computing: distributed sensor networks. J Vis Lang Comput 18(5):484–491
4. Mac Ruairí R, Keane MT, Coleman G (2008) A wireless sensor network application requirements taxonomy. In: Proceedings of the 2008 second international conference on sensor technologies and applications sensorcomm 2008, IEEE, No. 25–31 August, pp 209–216
5. García-Hernando A-B, Martínez-Ortega J-F, López-Navarro J-M, Prayati A, Redondo-López L (2008) Problem solving for wireless sensor networks. Springer, London, pp 177–209
6. Sreenan CJ, Silva JS, Wolf L, Eiras R, Voigt T, Roedig U, Vassiliou V, Hackenbroich G (2009) Performance control in wireless sensor networks: the ginseng project – [Global communications news letter]. IEEE Commun Mag 47(8):1–4
7. Antoniou M, Boon MC, Green PN, Green PR, York TA (2009) Wireless sensor networks for industrial processes. In: 2009 IEEE sensors applications symposium, vol 19, no. 6, pp 13–18
8. Shanmugaraj M, Prabakaran R, Dhulipala VRS (2011) Industrial utilization of wireless sensor networks. In: 2011 IEEE international conference on emerging trends in electrical and computer technology, pp 887–891
9. Suriyachai P, Brown J, Roedig U (2010) Time-critical data delivery in wireless sensor networks. Distrib Comput Sens Syst 6131:216–229

Appendices

Appendix A. Network Configuration: Example

In this Appendix, we show an example of how to define a network configuration.

Considering the network configuration represented in Fig. A.1, the user needs to convert it to a plain-text format to introduce it in the planning algorithm. Figure A.2 shows the corresponding plain-text format.

The plain-text format is defined by two structures: tree structure and treeAddr structure. The tree structure includes information about node connectivity. For each leaf node, it includes which nodes are part of the path between the leaf node and the sink node. The treeAddr structure maps the node id to a specific node address. This address depends on the network communication protocol. In this example we are using a hexadecimal addressing scheme.

J. Cecílio and P. Furtado, *Wireless Sensors in Industrial Time-Critical Environments*, Computer Communications and Networks, DOI 10.1007/978-3-319-02889-7,

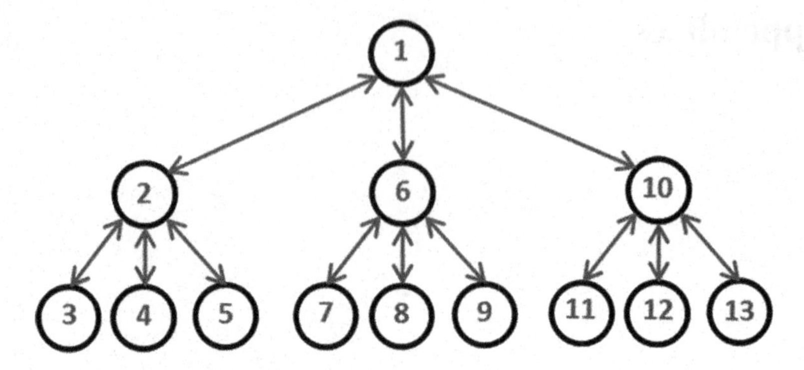

Fig. A.1 Network configuration

```
const Tree [MAX_LEAF_NODES] =         {
          {3, 2, 1},
          {4, 2, 1},
          {5, 2, 1},
          {7, 6, 1},
          {8, 6, 1},
          {9, 6, 1},
          {11, 10, 1},
          {12, 10, 1},
          {13, 10, 1}
};

const treeAddr[MAX_NODES] =        {
          {1, 0x0000},
          {2, 0x0002},
          {3, 0x0003},
          {4, 0x0004},
          {5, 0x0005},
          {6, 0x0006},
          {7, 0x0007},
          {8, 0x0008},
          {9, 0x0009},
          {10,0x000A},
          {11,0x000B},
          {12,0x000C},
          {13,0x000D}
};
```

Fig. A.2 Network configuration: plain-text format

Appendix B. Network Layout: Example

In this Appendix, we show an example of how to define a network layout to be recognized by the planning algorithm.

The network layout used includes all information about the network configuration (see appendix A) and a TDMA schedule. This schedule is defined by the user. It must be introduced in the algorithm in a plain-text fashion as shown in Fig. B.1.

```
const epoch[SLOTS_PER_EPOCH] = {

        //Upstream B1
        {0x0003,0x0002},
        {0x0003,0x0002},
        {0x0004,0x0002},
        {0x0004,0x0002},
        {0x0005,0x0002},
        {0x0005,0x0002},
        {0x0002,0x0000},
        {0x0002,0x0000},
        {0x0002,0x0000},
        {0x0002,0x0000},
        {0x0002,0x0000},
        {0x0002,0x0000},
        {0x0002,0x0000},
        {0x0002,0x0000},

        //Upstream B2
        {0x0007,0x0006},
        {0x0007,0x0006},
        {0x0008,0x0006},
        {0x0008,0x0006},
        {0x0009,0x0006},
        {0x0009,0x0006},
        {0x0006,0x0000},
        {0x0006,0x0000},
        {0x0006,0x0000},
        {0x0006,0x0000},
        {0x0006,0x0000},
        {0x0006,0x0000},
        {0x0006,0x0000},
        {0x0006,0x0000},

        //Upstream B3
        {0x000B,0x000A},
        {0x000B,0x000A},
        {0x000C,0x000A},
        {0x000C,0x000A},
        {0x000D,0x000A},
        {0x000D,0x000A},
        {0x000A,0x0000},
        {0x000A,0x0000},
        {0x000A,0x0000},
        {0x000A,0x0000},
        {0x000A,0x0000},
        {0x000A,0x0000},
        {0x000A,0x0000},
        {0x000A,0x0000},

        //Processing slots (= SLOTS_PROC)
        {0x0, 0x0},
        {0x0, 0x0},

        //TS (Time synchronization)
        {0x0000,0xffff},
        {0x0001,0xffff},
        {0x0002,0xffff},
        {0x0003,0xffff},

        //Downstream slots
        {0x0000,0xffff},
        {0x0002,0xffff},
        {0x0006,0xffff},
        {0x000A,0xffff}
};
```

Fig. B.1 TDMA schedule: plain-text format

Appendix C. Evaluation of Planning and Monitoring Approaches: Details

In this Appendix, we detail in form of tables the values shown in charts of Chap. 9 – an industrial case study. Each table includes in the caption the number of figures that the values are related to (Tables C.1, C.2, C.3, C.4, C.5 and C.6).

Table C.1 Monitor latency per node, Fig. 9.3

Node	Observed [avg]	Stdev	Observed [max]	Observed [P99]
2	33.68	0.55	44.70	36.05
3	113.69	0.52	124.04	115.96
4	173.90	0.45	178.30	176.06
5	173.99	0.90	188.07	176.65
6	33.52	0.48	43.86	35.49
7	114.07	0.55	124.10	116.52
8	173.96	0.63	184.58	176.40
9	173.63	0.55	185.73	175.95
10	33.68	0.52	47.28	35.94
11	113.63	0.50	124.16	115.91
12	173.58	0.69	187.07	175.83
13	173.91	0.60	184.58	176.20

Table C.2 Monitor latency per level with forecast, Fig. 9.4

Level	Observed [avg]	Stdev	Observed [max]	Minimum	Forecast [max]
Level 1	33.68	0.55	44.70	12.95	51.78
Level 2	113.69	0.52	124.04	92.91	131.78
Level 3	173.99	0.90	188.07	153.01	191.78

Table C.3 Monitor latency per network part, Fig. 9.5

Network part	Observed [avg]	Stdev	Observed [max]	Minimum	Forecast [max]
WSN	170.06	0.78	180.00	170.00	180.00
Serial	2.60	0.36	7.34	1.98	7.79
Middleware	1.33	0.28	3.33	0.67	3.99
Total	173.99	0.90	188.07	173.01	191.78

Table C.4 Event latency per node with forecast, Fig. 9.9

Node	Observed [avg]	Stdev	Observed [max]	Observed [P99]
2	33.68	0.55	44.70	36.05
3	113.69	0.52	124.04	115.96
4	173.90	0.45	178.30	176.06
5	173.99	0.90	188.07	176.65
6	33.52	0.48	43.86	35.49
7	114.07	0.55	124.10	116.52
8	173.96	0.63	184.58	176.40
9	173.63	0.55	185.73	175.95
10	33.68	0.52	47.28	35.94
11	113.63	0.50	124.16	115.91
12	173.58	0.69	187.07	175.83
13	173.91	0.60	184.58	176.20

Table C.5 Asynchronous closed-loop latency for all configurations, Fig. 9.14

	Case 1	Case 2	Case 3	Case 4
Observed (avg)	270.20	60.60	70.40	270.50
Stdev	0.31	0.15	0.22	0.34
Observed (max)	280.00	60.00	80.00	280.00
Forecast [max] if catches the next downstream slot	280.00	60.00	80.00	280.00
Forecast [max] if catches the second next downstream slot	380.00	160.00	180.00	380.00
User requirement	500.00	250.00	150.00	500.00

Table C.6 Synchronous closed-loop latency for all configurations, Fig. 9.15

	Case 1	Case 2	Case 3	Case 4
Observed (Avg)	55.0	57.6	52.3	54.3
Stdev	28.8	29.4	28.6	29.4
Observed (Max)	114.8	128.7	106.8	113.0
Forecast (Max)	131.8	131.8	131.8	131.8
User requirement	500.0	250.0	150.0	500.0

Index

J. Cecílio and P. Furtado, *Wireless Sensors in Industrial Time-Critical Environments*, Computer Communications and Networks, DOI 10.1007/978-3-319-02889-7,

W

Z

Printed in the United States
By Bookmasters